Haunted Soundtracks

New Approaches to Sound, Music, and Media

Series Editors: Carol Vernallis, Holly Rogers, and Lisa Perrott

Forthcoming Titles:

Musical New Media by Nicola Dibben
David Bowie and the Expansion of Music Video by Lisa Perrott
Popular Music, Race, and Media since 9/11 by Nabeel Zuberi
Bellini on Stage and Screen edited by Emilio Sala, Graziella Seminara, and Emanuele Senici
Traveling Music Videos edited by Tomáš Jirsa and Mathias Bonde Korsgaard
Kahlil Joseph's Transmedia Works: The Audiovisual Atlantic by Joe Jackson

Published Titles:

Transmedia Directors by Carol Vernallis, Holly Rogers, and Lisa Perrott
Dangerous Mediations by Áine Mangaoang
Resonant Matter by Lutz Koepnick
Cybermedia: Science, Sound, and Vision edited by Carol Vernallis, Holly Rogers, Jonathan Leal, Selmin Kara
The Rhythm Image: Music Videos in Time by Steven Shaviro
YouTube and Music: Cyberculture and Everyday Life edited by Holly Rogers, Joana Freitas, and João Francisco Porfírio
More Than Illustrated Music: Aesthetics of Hybrid Media Between Pop, Art and Video edited by Elfi Vomberg and Kathrin Dreckmann
Remediating Sound: Repeatable Culture, YouTube and Music edited by Holly Rogers, Joana Freitas, and João Francisco Porfírio
David Bowie and the Art of Music Video by Lisa Perrott
Haunted Soundtracks: Audiovisual Cultures of Memory, Landscape and Sound edited by K. J. Donnelly and Aimee Mollaghan

Haunted Soundtracks

Audiovisual Cultures of Memory, Landscape, and Sound

Edited by

K. J. Donnelly and Aimee Mollaghan

BLOOMSBURY ACADEMIC
NEW YORK • LONDON • OXFORD • NEW DELHI • SYDNEY

BLOOMSBURY ACADEMIC

Bloomsbury Publishing Inc, 1385 Broadway, New York, NY 10018, USA
Bloomsbury Publishing Plc, 50 Bedford Square, London, WC1B 3DP, UK
Bloomsbury Publishing Ireland, 29 Earlsfort Terrace, Dublin 2, D02 AY28, Ireland

BLOOMSBURY, BLOOMSBURY ACADEMIC and the Diana logo
are trademarks of Bloomsbury Publishing Plc

First published in the United States of America 2024
This paperback edition published 2023

Copyright © K. J. Donnelly and Aimee Mollaghan, 2024
Each chapter copyright © by the contributor, 2024

For legal purposes the Acknowledgments on p. x constitute
an extension of this copyright page.

Cover design: Louise Dugdale
Cover image: David Wall/Getty Images

All rights reserved. No part of this publication may be: i) reproduced or transmitted in any form, electronic or mechanical, including photocopying, recording or by means of any information storage or retrieval system without prior permission in writing from the publishers; or ii) used or reproduced in any way for the training, development or operation of artificial intelligence (AI) technologies, including generative AI technologies. The rights holders expressly reserve this publication from the text and data mining exception as per Article 4(3) of the Digital Single Market Directive (EU) 2019/790.

Bloomsbury Publishing Inc does not have any control over, or responsibility for, any third-party websites referred to or in this book. All internet addresses given in this book were correct at the time of going to press. The author and publisher regret any inconvenience caused if addresses have changed or sites have ceased to exist, but can accept no responsibility for any such changes.

Library of Congress Cataloging-in-Publication Data
Names: Donnelly, K. J. (Kevin J.) editor. | Mollaghan, Aimee, editor.
Title: Haunted soundtracks : audiovisual cultures of memory, landscape, and sound / edited by K.J. Donnelly and Aimee Mollaghan.
Description: New York : Bloomsbury Academic, 2023. | Includes bibliographical references and index. | Summary: "An examination of the relationship between audiovisual soundtracks and the "sonic haunting" of trauma, anxiety, and nostalgia"– Provided by publisher.
Identifiers: LCCN 2023013061 (print) | LCCN 2023013062 (ebook) | ISBN 9781501389559 (hardback) | ISBN 9781501389597 (paperback) | ISBN 9781501389573 (pdf) | ISBN 9781501389566 (epub) | ISBN 9781501389580 (ebook other)
Subjects: LCSH: Film soundtracks–History and criticism. | Motion picture music–History and criticism. | Haunted house films–History and criticism. | Landscapes in motion pictures. | Nostalgia in motion pictures. | Motion pictures–Sound effects.
Classification: LCC ML2075 .H37 2023 (print) | LCC ML2075 (ebook) | DDC 781.5/42–dc23/eng/20230524
LC record available at https://lccn.loc.gov/2023013061
LC ebook record available at https://lccn.loc.gov/2023013062

ISBN: HB: 978-1-5013-8955-9
PB: 978-1-5013-8959-7
ePDF: 978-1-5013-8957-3
eBook: 978-1-5013-8956-6

Series: New Approaches to Sound, Music, and Media

Typeset by Integra Software Services Pvt. Ltd.

For product safety related questions contact productsafety@bloomsbury.com.

To find out more about our authors and books visit www.bloomsbury.com
and sign up for our newsletters.

For Danijela

Contents

List of Illustrations	viii
Acknowledgments	x

1	Introduction *K. J. Donnelly and Aimee Mollaghan*	1
2	What wind in what trees? Listening to *Blow-Up* (1966) *Paul Newland*	9
3	Imaginal Space and the Occult Soundtrack in Guy Maddin's *Keyhole* (2011) *Daniel Bishop*	25
4	The Haunted and the Medium *K. J. Donnelly*	41
5	Producing Paranormal Sounds: Electronic Music, Projection, and Blurred Boundaries in *The Legend of Hell House* (1973) and *The Stone Tape* (1972) *Jamie Sexton*	57
6	Cries and Whispers: Landscape and Sound in *The Owl Service* (1969) and *Red Shift* (1978) *Craig Wallace*	73
7	Concrète Spaces: Musique Concrète in Gus Van Sant's *Paranoid Park* (2007) *Jessica Shine*	89
8	Haunted by Extinction: Sounding an Arctic Uncanny *Lisa Coulthard*	105
9	Acoustic Ghosts and Haunted Landscapes in Contemporary British Landscape Cinema *Aimee Mollaghan*	123
10	The Long Trajectory of Death: Justin Kurzel's Screen Adaptation of *Macbeth* (2015) *Danijela Kulezic-Wilson*	135
11	Haunted Folk: Specters of the Analogue in *Annihilation* (2018) *John McGrath*	151
12	Sonic Novelty and Conceptual Obscurity: Music, Landscape and Enigma in *Picnic at Hanging Rock* (Peter Weir, 1975) *Jady Jiang*	165
13	Outside Inside: Nature, Gender, and the Altered Domestic Space in *Possum* (1997) and *Nature's Way* (2006) *Andrea Wright*	185

Contributors	197
Index	200

List of Illustrations

2.1	Thomas enters the "park" space from *Blow-Up* (Carlo Ponti Productions, 1966)	12
2.2	Maryon Park as theatrical and sonic space from *Blow-Up* (Carlo Ponti Productions, 1966)	14
2.3	The sound of the shutter? Thomas capturing reality from *Blow-Up* (Carlo Ponti Productions, 1966)	15
3.1	(0:32:04) from *Keyhole* (Monterey Media, 2012)	30
3.2	(1:25:28) from *Keyhole* (Monterey Media, 2012)	33
3.3	(2:08:34) from *To Kill a Mockingbird* (Universal Home Video, 2006)	34
3.4	(1:29:52) from *Keyhole* (Monterey Media, 2012)	35
11.1	Cells Divide in "Official Trailer," *Annihilation* (Paramount Pictures, 2018)	156
11.2	Author transcription of "Cells Divide" alternate fingerpicking pattern excerpt	157
11.3	Kane returns from *Annihilation* (Paramount Pictures, 2018)	158
11.4	The crew approach "The Shimmer" from *Annihilation* (Paramount Pictures, 2018)	159
12.1	Diegetic classical music: the string quartet contrasts with the background landscape, with the pot plants demonstrating a mastery of cultivation in *Picnic at Hanging Rock* (British Empire Films Australia, 1975)	168
12.2 & 12.3	A girl appears and disappears in a cave in *Picnic at Hanging Rock* (British Empire Films Australia, 1975)	169
12.4	Sara's corpse in the garden in *Picnic at Hanging Rock* (British Empire Films Australia, 1975)	171
12.5	Disturbing landscape in *Picnic at Hanging Rock* (British Empire Films Australia, 1975)	172

12.6	Animals in the disturbing landscape in *Picnic at Hanging Rock* (British Empire Films Australia, 1975)	172
12.7	The girls fastening each other's dresses in *Picnic at Hanging Rock* (British Empire Films Australia, 1975)	177

Acknowledgments

The editors would like to thank Leah Babb-Rosenfeld and Rachel Moore at Bloomsbury Publishing for their kindness and support throughout this project. We would also like to thank our contributors for their patience and perseverance. This book has been incubating for a number of years and we are grateful that so many people were prepared to stick with it, particularly as the Covid pandemic disrupted all of our lives. Their work makes a significant contribution to contemporary debates on the shifting relationship between landscape, memory, and sound in audiovisual media. Finally, we would like to acknowledge and pay tribute to the scholarship and memory of our dear friend and contributor Daniela Kulezic-Wilson, who sadly passed away while we were working on this collection. She leaves behind a body of astounding work that will continue to have resonance for generations of scholars of music, sound, and the moving image.

1

Introduction

K. J. Donnelly and Aimee Mollaghan

In recent years, there has not only been an emerging interest in soundtracks in audiovisual culture but also an interest in the less solid spectral aspects of culture more generally. The turn of the Millennium has heralded a significant outgrowth of culture that demonstrates an awareness of the ephemeral nature of history and the complexity underpinning the relationship between location and the past. This has been especially apparent in the contemplation of the shifting relationship between landscape, memory, and sound in film, television, and beyond. The scope of inquiry for this collection of essays emphasizes the ineffable ghostly qualities of a certain body of soundtracks, extending beyond merely the idea of "scary films" or "haunted houses." Rather, the sonic haunting under consideration here is tied to trauma, anxiety, or nostalgia associated with spatial and temporal dislocation in the face of population moves and pressures, ecological issues and destruction of the traditional countryside, unstable borders, and porous boundaries, as well as increasingly intensified tourism and travel to consume landscape and location. This shift in focus from how audiovisual landscapes can be experienced from one of seeing to one of listening allows for an examination of how music and sound are essential to the composition of intangible topologies present in the texts investigated here. Rather than using landscape merely as a physical space for the site of action these representations of landscapes *haunted* by sonic ghosts allow for a psychological engagement with these sonically constructed landscapes.

The term "landscape" can be understood in a myriad of ways. As a culturally constructed concept, it implies a physical environment that is composed or manufactured, a shaping of the natural environment, something ideological that potentially informs the manner in which the world can be seen or experienced. Geographers such as James Duncan, Nancy Duncan, and Denis Cosgrove position landscape as a socio-political concept writing that the history of landscape can only be understood as part of a wider history of economy and society.[1] It has its own implications that represent the way certain groups of people have signified themselves and their world through their imagined relationship with nature, and through which they have underlined and communicated their own social role and that of others with respect to the external environment. As James Duncan asserts, landscape acts as a "signifying system through which a social system is communicated, reproduced, experienced, and explored."[2] If landscape is something that can be fabricated, then it is something that can be subject to multiple readings or interpretations particularly when encapsulated as an audiovisual artifact.

The audiovisual landscapes of interest here are interior and exterior, shaped and haunted not necessarily by spectral visibility or invisibility of image, but rather by sound. The diffuse nature of sound makes it a particularly useful medium for the presentation of trauma and nostalgia, providing a conduit for specters to whisper their secrets. To paraphrase Jacques Derrida, haunting is historical, but it is not dated.[3] It does not have a specific chronology or time. It functions outside of temporality, outside of linear narrative. It also has the ability to shift chronologies off-kilter. Due to the ability of sound to diffuse across space, it can bleed into or haunt an environment beyond what can visually be perceived, dissolving the discrete boundaries of the landscape. Sound can shape the spatial contours of the landscape by coalescing sonic ecologies. Ubiquitous, yet often largely unnoticed, these sonic or musical ecologies can help to construct aspects of cultural and personal identity. Furthermore, sound can blur the barriers between interiority and exteriority. It can detect aspects of the landscape not visible to the eye; geophones can allow us to hear sounds underneath the ground, hydrophones return subaqueous sounds from beneath the waves, ultrasound transducers shift the interior bruit of our bodies to the outside.

Acoustic ecologist, R. Murray Schafer, writes that "hearing is a way of touching at a distance,"[4] but more than that one could surmise that it is also a way of touching across time. Sound space is not the same as image space and differences in the time that it takes a sound to arrive at the ear tell us something about spatial relationships within an environment. Sonic phantoms manifest themselves through their aurality, their voices lurking out of time, diffusing through the temporal membranes of past, present, and future to conjure up a sonic identity that is at once both construct and memory, and, just like Schafer's concept of the soundscape, is both actual and abstract. Nowhere is this more obvious than in the moving image, where the space between sound and images can conjure ghosts, evident at times when some films lose direct "synch."[5]

The fleeting nature of memories inherently bestows on them a spectral character and sounds are particularly effective as potent memory triggers particularly when connected to a specific space or place. Songs, musical compositions, and even certain sounds are able to directly access the recesses of our minds and sometimes uncertain feelings from long ago. In a different process, perhaps, music can also carry collective as well as personal memory, and indeed some cultures appear to use the sonic to address the trauma of the past to try to heal the present, or at the least to try to break the impasses of present social and cultural issues.

Landscape should also be conceived as a palimpsest that can be written over in order to create and recreate senses of identity. This relies on collective memories passed down through generations, but also can engage that which ought to be forgotten for one reason or another. Indeed, "haunting" can make for an unstable and questioning presence that upsets the dominant order by implication and suggestion rather than a direct force of conflict.

Part of the wave of the so-called "sonic turn" in theory and the study of culture, this collection will focus in particular on audiovisual forms that foreground landscape, sound, and memory. It will address how soundscapes become or are an intrinsic part of landscape, and how this can hold a meaning or emotion that is nearly tangible but not explicit. This is the first book of its sort that deals with the ephemeral but affecting soundtrack in audiovisual culture (films, television, and other arts) in the light of sound's persistent but ambiguous relationship to location. As such, it will offer a distinct and novel approach to analysis drawn from a range of disciplines including, but not limited to, film studies, sound studies, philosophy, and cultural geography. This mode of analysis will be informed by psychogeography (a desire to consider location and landscape as essentially emotional or psychological objects rather than merely backdrops for drama), and Jacques Derrida's notion of "hauntology,"[6] which promotes a less positivist approach to history and finds forgotten, spectral history present in the margins of culture. Hauntology works against unitary and linear understandings of the past, registering that all past elements remain, sometimes as significant "absent structuring" or trace memory in something else. Consequently, the past in the present can be rethought and understood anew. This book contends that a particularly strong way of achieving this is through a consideration of landscape's relationship to sound in audiovisual culture, helping to concretize more supernatural or evanescent concerns into aesthetic forms or artifacts that validate their existence.

One issue that we have had to contend with in the process of developing this collection is the insufficiency of language to express the psychological, temporal, and spatial processes of haunting at play in the audiovisual texts under consideration here. Terms such as "psychogeography" or "hauntology," although pervasive, can sometimes seem vague or subject to perceived malapropism, severed from their etymological roots. The essays in this collection go some way toward articulating a syntax for concepts such as these, lifting the veil between the effable and ineffable to reify these ideas. Just as Guy Debord conceived the portmanteau psychogeography and Jacques Derrida contrived the term "hauntology," authors within this collection also find themselves in the position where existing words and idioms are not always adequate for their needs.

In order to engage with the sonic spaces under consideration in Antonioni's *Blow-Up* (1966) for example, Paul Newland draws on Mikhail Bakhtin's literary concept of the chronotope, a concept which accounts for the unification of space and time. Newland, however, refashions this concept into that of the "phonotope," an audiovisual analogue, which more fully accounts for sound spaces in *Blow-Up*. Through a process of close listening to the fabricated sound spaces in the film which he refers to as a "blowing up of the sound space," he asserts that wind sounds in particular evoke a sense of an uncanny, otherworldly natural presence reacting to the events of the film, while simultaneously trying to communicate the alienation experienced by photographer Thomas (David Hemmings), the central protagonist of the film.

Both K. J. Donnelly and Daniel Bishop examine the ambiguous relationship between space and the occult within a certain body of moving image. Donnelly interrogates how the soundtrack in Robert Eggers folk horror film *The VVitch* (2015) provides material form for a supernatural presence, haunting the landscape, and manifesting affective atmospheres. He identifies that the seemingly unstructured music in the film serves a predominantly spatial rather than temporal role, providing a dimensional rather than chronological continuum. Although visual and narrative structure helps to provide some temporal scaffolding for the sonic structure of the film, Donnelly suggests that the space in the film is constructed through the melding of the sound with the landscape, haunting the landscape and imbuing it with a sense of uncanniness.

Bishop makes a compelling case for the usefulness of philosopher Gaston Bachelard's phenomenological theories of space when probing the mutability of audiovisual spaces in screen media. Drawing on Bachelard's poetics of space, his chapter questions how Canadian filmmaker Guy Maddin's film *Keyhole* (2011) scrutinizes the connection between ghosts and the illusory physical space of the occult imagination. He posits that the score hovers somewhere between perceptible musical presence and amorphous, yet dynamic ambience. It is this spatial and material indistinctness which allows the soundtrack to diffuse across spatial and temporal boundaries and states.

At the core of this collection then, is a consideration of the porousness of boundaries. Jessica Shine demonstrates how the use of *musique concrète* can disrupt the traditionally delineated sonic spaces of diegetic and non-diegetic within the film soundtrack. She convincingly argues the case for extending Gus Van Sant's *Death Trilogy* (2002–5)[7] to a *Death Quartet* by including *Paranoid Park* (2007) amongst its corpus. Perhaps more importantly though, she contends that *musique concrète* haunts the diegetic space of the films of the *Death Quartet*. Expressly focusing on *Paranoid Park*, she questions how *musique concrète* can challenge our preconceptions of what a moving score might entail and how interior trauma might be expressed, in its ability to cross the "fantastical gap"[8] between the realm of the diegetic and non-diegetic.

Jamie Sexton's chapter explores how cult British films of the 1970s such as *The Stone Tape* (Sasdy, 1972) and *The Legend of Hell House* (Hough, 1973) utilize electronic sound produced by the BBC Radiophonic workshop to signify ghostly and occultic phenomena. He investigates how these films use electronic soundtracks to engage with science and the paranormal, yet simultaneously blur any such divisions. On a macro scale he extends this consideration of the sonic practices apparent in these moving image texts to interrogate how they destabilize temporal boundaries, creating ambiguities between where the sound design ends, and the score begins.

This is something that Craig Wallace also explores in his consideration of the relationship between landscape and sound in the British television adaptations of author Alan Garner's novels *The Owl Service* (ITV, 1967) and *Red Shift* (BBC, 1973). As Wallace asserts sound disrupts chronological distance and allows the stratified landscapes to be haunted by ghosts of the past, present, and future simultaneously. In *Red Shift* characters from different time periods seem to inhabit the Cheshire landscape concurrently. In a similar fashion, linear time and narrative unravel in the valley of *The Owl* Service, sonically haunted by traumatized ghosts of the present.

Many of the chapters in this collection center on sonic representations of trauma and grief. Aimee Mollaghan's chapter examines sonic haunting in British period dramas which have emerged since the beginning of the twenty-first century. Often incorporating Gothic or supernatural elements, these adaptations of nineteenth- and early twentieth-century novels locate themselves within the British rural landscape, irrespective of the original setting. The psychogeographical landscapes of trauma within the two case studies discussed in the chapter, *Wuthering Heights* (Andrea Arnold, 2011) and *Sunset Song* (Terence Davies, 2015), are constructed through a process of sonic haunting by spectral presences unconstrained by spatio-temporal borders.

Danijela Kulezic-Wilson's chapter on the "long trajectory of death" in Justin Kurzel's feature film adaption of *Macbeth* interrogates how the landscape embodies the memories of violent deaths, which psychologically haunt the character of Macbeth. Furthermore, she considers how the score both gives corporeal form to the forces of trauma and grief within the film which haunt these scarred landscapes, while also grieving for the characters.

In a similar fashion, John McGrath's chapter explores how Alex Garland's science fiction film *Annihilation* (2018) is haunted not only by its characters' traumatic histories, but also through the spectral traces of analogue technology. He proposes that the notion of analogue technology becomes a referent for more metaphysical themes connected to the dysregulated tendencies of humanity. In the world of the film, the protagonists are trapped in a cycle of repetition, absorption, and mutation in an uncanny shimmer world of *doppelgängers*. The crackle and materiality of the analogue score and folk music recordings haunt both the "weird" landscapes and the digitally manipulated sound of the film, imprinting them with a sense of human presence.

Turning these sonic manifestations of grief and trauma toward the Arctic, Lisa Coulthard's chapter considers what she refers to as "uncanny sonic geographies of extinction" in the Artic as expressed in two limited "polar Gothic" television series *The Terror: Season 1* (2018) and *The North Water* (2021). Rather than simply reviving or recreating sounds from the mid-nineteenth century, she avers that the landscapes of both series are sonically haunted by both an awareness of their own history, and a recognition of the present and future erosion of the polar ice caps.

Both Andrea Wright and Jady Jiang grapple with the Australian and New Zealand Gothic. Jiang uses Peter Weir's 1975 film *Picnic at Hanging Rock* as a way to interrogate national anxieties surrounding Australia's colonial past. Jiang explores the tensions between the civilized sites of colonial Victorian society and the untamed landscapes of the Australian outback. For her, the dichotomy of these spaces is sonically marked by a similar musical duality in which the classical music imposes a European culture on the landscape, whereas the idiosyncratic exoticism of the panpipes haunts the undomesticated spaces of the Australian wilderness.

Like Jiang, Wright probes the duality present in the representation of landscape and nature in Australian and New Zealand cinema. She also highlights the two readings of the Romanticized New Zealand landscape, which have become central to how it is presented on screen; it is either wild and sublime or cultivated and topographically well behaved. She locates the short films *Possum* (Brad McGann, 1997) and *Nature's Way* (Jane Shearer, 2006) within this tradition, questioning how the encroachment of uncanny and often threatening soundscapes from exterior to domestic spaces disrupt the settler myth of mastering nature.

Notes

1. See Denis Cosgrove, *Social Formation and Symbolic Landscape* (Madison: University of Wisconsin Press, 1984); James Duncan, *The City as Text: The Politics of Landscape Interpretation in the Kandyan Kingdom* (Cambridge: Cambridge University Press, 1990); James Duncan and Nancy Duncan, *Landscapes of Privilege: The Politics of the Aesthetic in an American Suburb* (Cambridge: Cambridge University Press, 2004).
2. Duncan, *The City as Text*, 17.
3. Jacques Derrida, *Specters of Marx: The State of the Debt, the Work of Mourning, and the New International* (London: Routledge, 1994).
4. R. Murray Schafer, *The Soundscape: Our Sonic Environment and the Tuning of the World* (Rochester, Vermont: Destiny Books, 1977/1994), 11.
5. K. J. Donnelly, *Occult Aesthetics* (Oxford: Oxford University Press, 2013).
6. Derrida, *Specters of Marx*.
7. Gus Van Sant's *Death Trilogy* consists of the following three films: *Gerry* (2002), *Elephant* (2003) and *Last Days* (2005).
8. Robynn J. Stilwell, "The Fantastical Gap between Diegetic and Nondiegetic" [in] Beyond the Soundtrack: Representing Music in Cinema," in *Beyond the Soundtrack: Representing Music in Cinema*, ed. Daniel Goldmark, Lawrence Kramer and Richard D. Leppert (Berkeley: University of California Press, 2007), 186.

2

What wind in what trees?
Listening to *Blow-Up* (1966)

Paul Newland

In the last interview he gave before his death in 1948, the American filmmaking pioneer D. W. Griffith said "What the modern movie lacks is beauty—the beauty of moving wind in the trees, the little movement in a beautiful blowing on the blossoms in the trees."[1] Moving images of natural phenomena such as wind moving tree branches enthralled early cinema audiences. Siegfried Kracauer described such examples of the magic of cinema thus: "undulating waves, moving clouds, and changing facial expressions [...] conveyed the longing for an instrument which could capture the slightest incidents of the world about us [...] whose incalculable movements resemble, somehow, those of waves or leaves."[2] In the work of Kracauer, wind in the trees "is taken to reveal cinema's ability to show the autonomy of the world unfold independently of authorial control."[3] Perhaps Griffith would have appreciated the memorable sequences of trees blowing in the wind in Michelangelo Antonioni's film *Blow-Up* (1966), one of three films that the Italian director was contracted to make in English for MGM. Produced by Carlo Ponti, *Blow-Up* won the *Palme d'Or* at Cannes in 1967.[4] The film stars David Hemmings as a London-based photographer Thom who, after taking a series of photographs in a suburban park, subsequently prints and blows up these images to discover what he believes to be evidence of a murder.

Sequences in *Blow-Up* shot in Maryon Park, south-east London, and a linked sequence in which the photographer Thomas (Hemmings) prints and blows up photographs he takes in this park, have received several critical interpretations. None of this criticism has focused primarily on Antonioni's use of sound. That is what I aim to address here. Iain Sinclair describes Maryon Park as an "amphitheatre, a wooded bowl, with tennis courts at the centre."[5] The art director Assheton Gordon chose this park as a setting for the "murder," because it was a "theatre box" and "it resembled the spatialities of de Chicoro, an artist admired by himself and Antonioni."[6] In this chapter, I will focus specifically on the sonic representation of leaves blowing in trees in this park in order to argue that Antonioni's carefully constructed sound space in *Blow-Up* purposefully avoids depicting quotidian or natural sounds in any straightforwardly realistic or naturalistic way.[7] Rather, I will show through close listening, or through a critical "blowing up" of the sound space, that sound in *Blow-Up* is constructed alongside the images to develop an aesthetic spatiality which achieves two key things. Firstly, the sound of wind in the trees can be interpreted as an evocation of the uncanny, almost supernatural reaction of nature to the dramatic proceedings of the film. As such, sound becomes suggestive of the trees as a kind of protagonist, as a meta-human, dramatic presence in the narrative. In this way, the sound of wind in the trees effectively operates as a sonic example of pathetic fallacy. Secondly, at the same time, I will show that the sound of the wind in the trees serves to evoke or communicate the psychological world of the key protagonist Thomas, primarily to communicate his condition as a profoundly dislocated and alienated modern figure.[8] I will show that the cumulative effect of these potential representational meanings of the sounds in the film constructs an existential and uncanny spatial framework which facilitates the communication of the constructed nature of cinematic representation, and through this, the constructed nature of human experience, and, more broadly, of human experience of modern reality itself.[9]

In order to analyze the sound space of *Blow-Up*, I want to draw on and further develop my concept of the "phonotope" (the prefix "phono" deriving from the Greek work for "sound"), which might best be understood as an audio-visual redevelopment of Mikhail Bakhtin's notion of the literary chronotope.[10] For Bakhtin, a chronotope is a "time-space."[11] Bakhtin was concerned with how the literary form of the novel produces chronotopes, writing that "every literary image is chronotopic."[12] The phonotope might best be understood as a "sound-space" which temporally informs and structures the spatial imaginary and, as it does this, transcends material, "real" places represented by the images in a film. In other words, I conceive of phonotopes as filmic time-spaces in which film sound aesthetics employed in representing real places develop a complex spatial and temporal dialogue with these "real" places. Just as Thomas blows up a series of images in the film in order to uncover the potential outcome of a series of events, I will "blow up" these film sounds in order to demonstrate the extraordinary representational complexity of Antonioni's constructed phonotope. In terms of methodology, this sonic "blowing up," or close listening, echoes aspects of "reduced listening." Michel Chion points out that "reduced listening does not forbid listening otherwise [...] reduced listening overlaps with the others and enriches them."[13] I am however advocating for a "spatial listening" that might facilitate an awareness of how and why filmic phonotopes are constructed, and how they convey meaning.

Antonioni's films are profoundly spatial. Writing in 1975, the film critic Penelope Houston advocated that Antonioni's greatest gift as a filmmaker was "his hypersensitive feeling about places, and the part landscape plays in mood."[14] But landscapes in Antonioni's films are sonic as well as visual. From the 1960s onwards, Antonioni's films often featured environmental sounds to help construct this spatiality.[15] *Blow-Up* develops a complex representational sound space (phonotope) which serves to evoke the quotidian life of not just specific places in London while at the same time rendering these places uncanny through aesthetic gestures toward their representational artifice. The sound for *Blow-Up* was recorded live on location, but Antonioni resisted simply synchronizing this sound to his images. The director has explained the importance of the process of sound design in his films thus: "My rule is always the same: for each scene, I record a soundtrack without actors."[16] He chose to meticulously plan and structure his sound world *after* shooting sequences.[17] Exploring the sequences shot in the park, I will now demonstrate how this structured, constructed sound world operates.

The First Visit to Maryon Park

Thomas's first visit to Maryon Park occurs when he drifts into this place after exploring a nearby antiques shop that he is considering purchasing. As William Arrowsmith puts it, "Of his own choice, he freely consents to the pull of the park, following his eyes—not his camera—where they lead him, entering slowly, even gravely, into the world opening out before him."[18] The entry into the park "marks a point of transition, for Thomas is carrying his old assumptions of power into an environment where they no longer obtain."[19] But this is also a point of transition in the film's employment of sound: the moment when sound starts to become more obviously artificial, and, as such, representational.

At the beginning of this sequence, Thomas is pictured photographing the antiques shop, with the entrance to the park behind him, tall trees blowing in

Figure 2.1 Thomas enters the "park" space from *Blow-Up* (Carlo Ponti Productions, 1966).

the wind. The film then cuts to a shot from a camera within the park, looking back at Thomas and the shop, framed by the trees. The gentle sound of the wind blowing the leaves increases in volume now, while the camera remains static. Peter Brunette advocates that these "wind in the trees" sounds are invested in a "foreboding and existential resonance."[20] At first, this sound of the wind in the trees appears to be an "ambient sound"; the type of sound, as Michel Chion explains, "that envelops a scene and inhabits its space, without raising the question of the identification of visual embodiment of its source: birds singing, churchbells ringing. We might also call them territory sounds, because they serve to identify a particular locale through their pervasive and continuous presence."[21] However, the sound of the wind in the trees in this sequence gradually develops beyond any simple background ambience. Instead, this sound comes to operate as a hauntingly expressive representational device. It crucially informs the strange spatiality in the sequence in Maryon Park, which on the one hand facilitates Thomas's material and psychological removal from the quotidian life of the city, while on the other hand immersing him in an uncanny space of mystery.

The film cuts to a shot of a woman in a suit and hat (a mysterious, incongruous Antonionian figure) picking up litter from the lawn and the path in the park. The camera slowly pans to the left and she notices Thomas, who walks toward the camera. The film then cuts to a shot of this area of the park from another angle. The camera slowly pans left to reveal tennis courts and a circular flower bed. Here the sounds of the leaves and the birdsong continue at approximately the same ambient volume level. But the movement of the camera and subsequently its sudden static position (moving toward and then holding the image of the flower bed and tennis court for several seconds in the same shot) invest the landscape with a sense of mystery. This mystery is heightened by the incongruence of the relationship between images and sound. In other words, the camera movement helps to invest the sonic ambience with a strangeness which intensifies as the sequence develops. This sonic ambience is more noticeable as time unfolds in the sequence, and it becomes increasingly clear that this sound is not communicating the life of a straightforwardly "real" place, but that it is instead serving to communicate and underline the representational aspects of the film images and film sound we are witnessing, and to articulate the existence of an uncanny agora that exists between recorded images, recorded sound, and the "real" world recorded. With the camera still static, Thomas walks off into the distance, past the tennis courts, toward some birds gathered on the distant park lawn. We hear the sound of a tennis ball being hit. At this moment, the film cuts to a medium close-up of Thomas. The leaves are quieter now, but the sound of the tennis ball being hit remains, at the same volume. The gentle sounds of the wind blowing the leaves, the tennis ball being hit, and the bird singing continue, communicating the construction of a strange ambient space, but also, at the same time, evoking a space of mystery and foreboding. Here, the sound of wind in the trees also starts to communicate the uncanny reaction of features of a "natural" environment to the dramatic proceedings of the film. This sound becomes increasingly suggestive of the trees as a meta-human presence, while at the same time facilitating the aesthetic communication of Thomas's existential experience.

Thomas raises his camera to his eye and focuses on something. The film now cuts to a long shot of Thomas chasing the birds on the vast lawn with his camera. The sound of the leaves is barely audible here, but we can hear a bird singing again. Another cut sees the camera panning quite rapidly and without smoothness from right to left, very briefly capturing two figures in the corner of the frame (more on these figures in my conclusion), before focusing on the rear elevation of terraced houses beyond the park. Antonioni cuts again to another medium shot of Thomas on the grass. The leaves pick up in volume slightly now. The length of this sequence, without dialogue or any obvious dramatic purpose, also serves to lend it an uncanny quality. It communicates a sense of time unfolding in a specific place, while at the same time problematizing the ontological nature of this place. The film now cuts to a couple making their way up a steep bank, toward some bushes. The sound of the leaves in the wind increases in volume here, again signaling increased tension and mysteriousness. As Thomas playfully runs up some steps toward a higher section of the park, the sonic landscape steadily moves beyond any straightforwardly realistic representation of a place, toward a more obviously discernible representational artifice.

As Thomas enters this elevated park space, the sound of the leaves grows louder still. Tension noticeably increases. Thomas stops at the top of the steps, and peers through a branch, his eyes framed by leaves. He puts his camera to his eye again, and looks around this upper level of the park, clearly intrigued by something. The film now cuts to a panning shot, the camera moving from left to right. The couple can now be seen standing on the grass in the distance, holding hands. In the background, beyond the trees and bushes that mark the boundary of the park, 1960s-style buildings are visible. The sound of the leaves being blown by the wind is now very prominent. We hear the woman laugh. In the next shot, we see Thomas slowly climb over a low fence, and begin to photograph the couple. The sound of the leaves continues. A dog barks in the distance. The shutter on Thomas's camera can be heard rhythmically opening and closing: capturing reality. The volume of the leaves blowing shifts now: growing quieter, then louder again, coming in waves. It is clear that Antonioni wants us to *notice* this sound, for it to shape our experience of this mysterious sequence in this mysterious representational space.

Figure 2.2 Maryon Park as theatrical and sonic space from *Blow-Up* (Carlo Ponti Productions, 1966).

Figure 2.3 The sound of the shutter? Thomas capturing reality from *Blow-Up* (Carlo Ponti Productions, 1966).

Thomas hops over the low fence and runs toward a tree, and crouches and hides. He takes more photographs of the couple. With the new mechanical, rhythmic sounds of the shutter opening and closing on the camera, the volume of the leaves blowing increases again, also ratcheting up the tension while at the same time foregrounding and emphasizing the representational aspects of the sequence, and indicating the potential importance of the photographs being taken to what will unfold. Interestingly, the sound of the leaves swells in volume whenever the couple are pictured clearly in the frame, potentially communicating their importance to future events that will unfold. Michel Chion developed the term "anempathetic sound," which signifies a sound which appears to exhibit conspicuous indifference to what is going on in a film, while at the same time creating a sense of the tragic.[22] But this sound of the leaves in this sequence might effectively become an empathetic sound, a sound that seems to speak, like a response, or an intervention, to the events unfolding.

Noticing he has been spotted, Thomas turns and walks back toward the steps. The woman runs toward him, and the leaves noticeably swell in volume once more. Throughout the following sequence—during which the woman, Jane (Vanessa Redgrave) confronts Thomas about taking photographs of her and her lover—the leaves swell in volume whenever she is on screen, and quieten down again when she is offscreen. However, this pattern of sound changes after she bites Thomas's hand and he asks, aggressively, "What's the rush?" From now on the sound of the leaves remains more obviously constant throughout the sequence.

The last words Jane speaks on this initial meeting in the park are: "No we haven't met. You've never seen me." This is a key piece of dialogue, foreshadowing the series of disappearances that will occur later in the film. Antonioni gives us a wide shot of the park as she runs off behind a tree into the distance. The leaves swell in volume once more. When the film cuts again to a shot of Thomas walking back toward the antiques shop, the sound of the leaves suddenly disappears. Music appears instead (a slide guitar), which subsequent shots suggest might be diegetic, coming from a record player in the shop. This lack of a clear sound source, and the concomitant disruption between diegetic and extra-diegetic spaces, once again serves to communicate—or even foreground—the representational aspects of the film.

This first park sequence develops a highly complex representation that might initially be read through what R. Murray Schafer termed "hi-fi" soundscape: "one in which discrete sounds can be heard clearly because of the low ambient noise level."[23] Interestingly, Schafer's ecological perspective on "real" soundscapes led him to argue that the countryside is "generally more hi-fi than the city; night more than day; ancient times more than modern. In the hi-fi soundscape, sounds overlap less frequently; there is perspective-foreground and background."[24] Rather, in this sequence in *Blow-Up*—as in many Antonioni films—the representation of an outdoor place develops a quiet, "hi-fi" soundscape, which lends this space mysteriously uncanny, almost supernatural status.

The sound of the leaves blowing in the wind in this sequence might also be read as the articulation of the ghostly voice of a nonhuman sound source. Theoretical debates about the nonhuman in recent years have been driven by writers such as Jane Bennett and Eduardo Kohn.[25] Other writers have recently developed theories of sound which begin to engage with such theories of the nonhuman and anthropomorphism germane to my argument. For example, Mark Grimshaw and Tom Garner argue that sound is "emergent perception"; that sound should not be theorized as "object," but instead is constituted through plural material and immaterial mediations, as what they call a "sonic aggregate."[26] This concept of the sonic aggregate, if considered spatially, might allow us to understand how far the sound space of *Blow-Up* toys with or purposefully disrupts our previous knowledge (conscious or unconscious) of the relationship between sounds and their sources, and problematizes what Grimshaw and Garner term the "virtual cloud of potentials" from which the sound as perception emerges. Furthermore, working on nonhuman sound, Georgia Born's view is that:

> Generally, nonhuman sound is not a focus of human attention. To become aware of it requires an attunement, a shift from perpetual background to foreground, whether it is high-volume environmental sound of the more continuous ebb and flow of low-level hushes, hums, washes, and clusters of sonic events (trees rustling, planes and trains passing, fridge humming, flies buzzing, house creaking, cars revving, birds calling, construction work proceeding, and so on). Nonhuman sound exists as a constant, potentially affect-laden companion to quotidian life.[27]

Born builds on these ideas, and on the work of Alfred Whitehead,[28] as she seeks to "open up a conceptual space in which we understand sound, including nonhuman sound, as an inherently relational and 'mediational' phenomenon that overcomes dualistic understandings of subject and object and that [...] itself participates in subjectivity."[29] These ideas allow us to consider how far Antonioni's *Blow-Up* disrupts the "relational" and "mediational" aspects of sound in order to explore existential subjectivity.

In *Blow-Up*, this space opened up by the tension between the human and unhuman, reality and its cinematic representation ultimately produces the effect of a haunting, which chimes with the "murder" narrative of the film. As such, the sound space in the film can be read through Mark Fisher's work on hauntology. Drawing on Jacques Derrida, Fisher argues that there are two directions in hauntology: "The first refers to that which is (in actuality is) no longer, but which is still effective as a virtuality (the traumatic 'compulsion to repeat,' a structure that repeats, a fatal pattern). The second refers to that which (in actuality) has not yet happened, but which is already effective in the virtual (an attractor, an anticipation shaping current behavior)."[30] Thus, "Haunting can be seen as intrinsically resistant to the contraction and homogenization of time and space. It happens when a place is stained by time, or when a particular place becomes the site for an encounter with broken time."[31] As such, "specific (hauntological) landscapes—landscapes stained by time, where time can only be experienced as broken, as a fatal repetition."[32] *Blow-Up* certainly presents a landscape "stained by time," which can only be experienced as "repetition." David Toop has written about the potentially spectral aspects of sound. For example, he advocates that:

> sound is a haunting, a ghost, a presence whose location in space is ambiguous and whose existence in time is transitory. The intangibility of time is uncanny—a

phenomenal presence both in the head, at its point of source and all around—so never entirely distinct from auditory hallucinations. The close listener is like a medium who draws out substance from that which is not entirely there.³³

Toop's view of the spectral nature of sound chimes with the employment of sound in Michelangelo Antonioni's films, not least in *Blow-Up*.

Blowing Up Photographs of Maryon Park

The sequence in which Thomas blows up his photographic "capturing of the landscape" can also be read through Mark Fisher's work on hauntology and Toop's thoughts on sound as haunting. We see Thomas developing his film exposures, and slowly and systematically examine these images, printing a selection of enlargements which he arranges in a sequence, tacking them up onto the walls of his studio. After developing further magnifications of key portions of two of the exposures, Thomas eventually believes he can see images of a gunman and a corpse. These blown-up up images are grainy, effectively as unreadable as his friend Bill's (John Castle) abstract paintings. As this sequence unfolds we hear the clear, seemingly diegetic sounds of Thomas's footsteps on the wooden floorboards of the studio as he displays these images. There is also an ambient sound audible here, which might or might not be the distant hum of city. Thomas puts a vinyl LP on his record player, and sits and looks at the prints. But when he starts to look at an image of the park with a magnifying glass, this "diegetic" music abruptly fades to silence. This sonic moment signifies a profound breakdown between reality and representation. At this precise moment in the film, any relationship between diegetic and extra-diegetic worlds falls apart. Here Antonioni is evidently foregrounding the fact that the film we are watching is artifice, a *representation* of reality, just like the photographs Thomas is examining. As Thomas places more and more pictures around the room, they begin to appear like the frames of a film, like a moving picture. As Antonioni shows us these black-and-white photographs in close up in the following sequence, the sound of leaves blowing in the wind in the park inexplicably returns. This, like the sudden drop out of the jazz music moments before, also serves to highlight and foreground the constructed nature of the representational aspects of what is being witnessed here. This is a sonic example of Derrida's "specter," an aesthetic device evidently resistant to the contraction and homogenization of time and space.

The Third Visit to Maryon Park

After discovering what he believes to be images of a corpse and a gunman, Thomas drives back to the park for a second time at night. He parks his Rolls Royce by the antiques shop and walks into the park, once again climbing the steps to the high grassy platform. The sound of the leaves appears once again here, but it is now noticeably louder in volume than during the earlier park sequences. Thomas walks across the lawn toward the spot where he saw the body the night before. But the volume of the leaves in the trees significantly decreases when he arrives at what he believes to be the precise location and as he looks down at the grass where the body had previously lain, before increasing in volume again as he crouches down, camera in hand. The film cuts to a shot of Thomas. We are looking down at his back as he crouches on the grass. He looks up, over his shoulder. The sound of the leaves in the wind noticeably increases in volume again here. The film then cuts to an image of leaves blowing in the wind. The wind sounds even louder now. The camera pans back to Thomas, who is now standing, looking bemused, the leaves behind him blowing hard in the wind. He moves forward slowly, and stops. A mysterious white sign high on scaffolding beyond the perimeter of the park is now visible behind the fence and the bushes, lit up, but out of focus. Antonioni cuts again to an image of the grass, a tree, and bushes behind, where the steps lead away. He then cuts back to Thomas in the same position as before. Thomas turns his head back to where the body once lay, and then back toward the steps. At this precise moment we see the light on the large white sign suddenly go out, and the sign coming into full focus. As Thomas notices the change in light out of the corner of his eye, he turns to look at the sign. This is a typical Antonionian device—an unreadable sign, a sign with no ultimate meaning (intended or otherwise), which nevertheless looms over the events in the park, and is evidently brought to our attention to in this moment.[34] The leaves are quieter again now, as Thomas looks at the white sign, but they subsequently pick up in volume once more, as Antonioni cuts to a shot of Thomas standing, once again looking at the spot where he had found the body. This shot—taken by a static camera—is held for 28 seconds. It allows us to view Thomas looking around this specific place in the park for one last time, as the sound of the leaves blowing in the trees continues, along with the sound of what once again appears to be a dog barking in the distance. This sequence further cements the construction of a haunted sound space that is informed by the tension between the human and unhuman, and between the sounds of natural, nonhuman material objects and their cinematic representation. Again, during this sequence the sound of the wind in the trees functions as dramatic device, articulating the uncanny reaction of meta-human nature, and evoking a hauntological space, while at the same time articulating the existential experience of Thomas.

What follows is the famous mimed tennis match sequence. By now, Thomas has abandoned the murder mystery.[35] And by now, Antonioni has abandoned any pretense to any straightforwardly realistic representation of reality. The tennis "match" starts in near silence. Gradually the sound of the leaves blowing fades up from silence, and develops in waves throughout, rising and lowering in volume. As a young woman and man "play" tennis with an imaginary ball, other sounds can be heard in the sound space (phonotope), which might or might not be their footsteps on the court (it is important to note that not all the footsteps that can we see here have obviously corresponding sounds on the soundtrack), the players making the sound of a tennis ball being hit by rackets, or a tennis ball itself being hit. After Thomas picks up and throws the imaginary ball back to the players on the court, the camera remains on him as his eyes follow the flight of the imaginary ball, moving from left to right as the "game" recommences. During this sequence, a sound of a tennis ball being hit is clearly audible on the sound track, as Thomas looks on. The sound of the leaves swells once more, before the film cuts to a final image of Thomas, in long shot, on the grass. Here, Hancock's upbeat jazz joins the sound space, before the image of Thomas fades to nothing, with only the patchy grass remaining. The disjunct between sounds and images of sound sources in the final mimed tennis match sequence in *Blow-Up* becomes distinctly hauntological, a representation of Derrida's "specter," an example of the visibility of the invisible.

Conclusion

As I noted at the beginning of this chapter, Siegfried Kracauer once argued that images of wind in trees "reveal cinema's ability to show the autonomy of the world unfold independently of authorial control."[36] But Antonioni once explained that in *Blow-up* he was really "questioning the nature of reality."[37] Instead, in this film, the sound of wind in the trees does not in any simple way unfold independently of authorial control, or capture natural events unfolding, but instead serves—through authorial control—to evoke the uncanny reaction of nature to dramatic proceedings. As such, sound becomes suggestive of the trees in the park as a meta-human presence, while at the same time facilitating an aesthetic evocation of the existential experience of Thomas.

I want to end with a very short auto-ethnographic intervention. Before researching this chapter, I had already viewed (and listened to) *Blow-Up* dozens of times. Despite the fact I had taught the film to masters students over a number of years, re-watching (and re-listening to) the film again in 2022, I noticed several things for the first time. For example, it occurred to me that the couple in the first sequence in the park can be seen (before we see the shot of the climbing the bank) for a very brief moment, as the camera pans from right to left, away from the tennis court and toward the white terraced houses. It also occurred to me that if I ever get the opportunity to view the film on a big screen and will thus be able to "blow-up" the image, so to speak, I might be able to see a corpse lying on the grass when Jane (Redgrave) runs away from Thomas (Hemmings) at the end of the first park sequence. There was also much on the soundtrack I noticed for the first time, only recently, by listening very carefully multiple times—critically "blowing up" the sound. It was only after re-viewing and re-listening to the film that I concluded that I can hear no sound of a gunshot in the first key sequence in Maryon Park, despite the relative quietness and emptiness of the park during these moments. On considering these surprising new discoveries it occurred to me that the shot in the film of Thomas with a large magnifying glass, looking at the film negatives in his studio, was effectively the same activity I was currently engaged in—trying to make sense of a text critically, to *find* something in it. This made me ask myself the following question: what am *I* viewing and listening to, what am *I* seeking in the film?

I came to the conclusion that Antonioni's *Blow-Up* is showing us that recorded sound is always open to interpretation, as all images are, and that *all* sound is uncanny, not only because of the space (and indeed time) between the source of a sound and our awareness and consciousness of it, but also because of the fact that we are only ever hearing and interpreting sound waves formed at a sound's source. The film articulates the ultimate impossibility of an artist to accurately capture or depict objective truth or reality. As Iain Foreman points out, "If we acknowledge that soundscapes estrange, dislocate, render uncanny—they absent community—what is the role of sound as a medium of communication?"[38] Chatman and Duncan argue that by the end of *Blow-Up*, Thomas "is no more real than the imaginary tennis ball that he threw back into the tennis court."[39] Antonioni effectively shows us that *any* film—its story, its characters, its location, and its sounds—is no more real than Thomas's imaginary tennis ball.

Notes

1. Ezra Goodman, *The Fifty-Year Decline and Fall of Hollywood* (New York: Simon and Schuster, 1961), 19. See also Daniel Fairfax, "The Beauty of Moving Wind in the Trees: Cinematic Presence and the Films of D.W. Griffith," in *A Companion to D.W. Griffith*, ed. Charlie Keil (Oxford: Wiley Blackwell, 2018), 74–105.
2. Siegfried Kracauer, *Theory of Film: The Redemption of Physical Reality* (Princeton, NJ: Princeton University Press, 1997), 27–8.
3. Jordan Schonig, "Contingent Motion: Rethinking the 'Wind in the Trees' in Early Cinema and CGI," *Discourse* 40, no.1 (2018): 30–61; 31.
4. Robin Gregory was the sound recordist on *Blow-Up*, while Mike Le Mare was the sound editor, and J.B. Smith the dubbing mixer. Other uncredited individuals worked on the film sound in various capacities, including W. Carr (sound assistant), Fernando Caso (sound effects editor), Arkadi De Rakoff (assistant sound), Alvaro Gramigna (Foley artist), Ray Palmer (sound assistant), and Michael Sale (sound assistant).
5. Iain Sinclair, *Lights Out for the Territory* (London: Granta, 1998), 347.
6. David Alan Mellor, "Fragments of an Unknowable Whole: Michelangelo Antonioni's Incorporation of Contemporary Visualities in London, 1966," *Visual Culture in Britain* 8, no.2 (2007): 45–61; 56.
7. Seymour Chatman, *The Surface of the World* (Berkeley: University of California Press, 1985), 134.
8. Peter Brunette, *The Films of Michelangelo Antonioni* (Cambridge: Cambridge University Press, 1998), 131–2.
9. David Forgacs, "In the Details," *Blow-Up* Blu-Ray, Criterion (2017), 12–25; 21.
10. I have previously developed the concept of the "phonotope" in "Folksploitation: Charting the Horrors of the British Folk Music Tradition in The Wicker Man (Robin Hardy, 1973)," in *British Cinema in the 1970s*, ed. Robert Shail (London: British Film Institute, 2008), 119–28, and in "The Spatial Politics of the Voice in Patrick Keiller's Robinson in Ruins," *The New Soundtrack* 6, no.2 (2016): 129–42.
11. Mikhail M. Bakhtin, "The Dialogic Imagination: Four Essays," in *Discourse in the Novel* ed. Michael Holquist, trans. Caryl Emerson and Michael Holquist (Austin and London: University of Texas Press, 1981), 84.
12. Bakhtin, "The Dialogic Imagination," 251.

13 Michel Chion, "Reflections on the Sound Object and Reduced Listening," in *Sound Objects*, ed. James A. Steintrager and Rey Chow (Durham and London: Duke University Press, 2019), 23–32; 31.

14 Penelope Houston, "Keeping Up with the Antonionis," *Sight and Sound* 33, no.4 (1964): 163–8; 166.

15 Andy Birtwistle, "Heavy Weather: Michelangelo Antonioni, Tsai Ming-liang, and the Poetics of Environmental Sound," *Quarterly Review of Film and Video* 32, no.1 (2014): 72–90; 74.

16 Betty Jeffries Demby and Larry Sturhahn, "Antonioni Discusses The Passenger," in *Michelangelo Antonioni Interviews*, ed. Bert Cardullo (Jackson: University Press of Mississippi, 2008), 104–14; 111.

17 Antonella C. Sisto, *Film Sound in Italy: Listening to the Screen* (New York: Palgrave Macmillan, 2014), 137.

18 William Arrowsmith, *Antonioni: The Poet of Images*, ed. Ted Perry (New York and Oxford: Oxford University Press, 1995), 115–16.

19 Rodney Stenning Edgecombe, "The Emblematic Texture of Antonioni's Blow-Up," *Film Criticism* 3, no.1 (Fall 2011): 68–84; 82.

20 Brunette, *The Films of Michelangelo Antonioni*, 116.

21 Michel Chion, *Audio-Vision: Sound on Screen*, ed. and trans. Claudia Gorbman (New York: Columbia University Press, 1994), 75.

22 Ibid., 221–2.

23 R. Murray Schafer, *The Soundscape: Our Sonic Environment and the Tuning of the World* (Rochester, Vermont: Destiny Books, 1994), 43.

24 Ibid.

25 See Jane Bennett, *Vibrant Matter: A Political Ecology of Things* (Durham: Duke University Press, 2010), Eduardo Kohn, *How Forests Think: Toward an Anthropology beyond the Human* (Berkeley: University of California Press, 2013).

26 Mark Grimshaw and Tom Garner, *Sonic Virtuality: Sounds as Emergent Perception* (Oxford: Oxford University Press, 2015), 166–78.

27 Georgia Born, "On Nonhuman Sound—Sound as Relation," in *Sound Objects*, ed. James A. Steintrager and Rey Chow (Durham and London: Duke University Press, 2019), 185–207; 188.

28 See for example Alfred Whitehead, *The Concept of Nature* (Cambridge: Cambridge University Press, 1995), and *Modes of Thought* (New York: Free Press, 1968).

29 Born, "On Nonhuman Sound—Sound as Relation," 198.

30 Mark Fisher, "What Is Hauntology?," *Film Quarterly* 66, no.1 (2012): 16–24; 19.

31 Ibid.

32 Ibid., 21.

33 David Toop, *Sinister Resonance: The Mediumship of the Listener* (London and New York: Continuum, 2011), xv.

34 Edgecombe, "The Emblematic Texture of Antonioni's Blow-Up," 68–84; 74.
35 Sam Rohdie, *Antonioni* (London: British Film Institute, 1990), 46.
36 Kracauer, *Theory of Film,* 27–8.
37 Michelangelo Antonioni, "Antonioni—English Style," in *Blow-Up, a Film by Michelangelo Antonioni* (New York: Simon and Schuster, 1971), 4.
38 Iain Foreman, "Uncanny Soundscapes: Towards an Inoperative Acoustic Community," *Organised Sound* 16, no.3 (2011): 264–71; 265.
39 Seymour Chatman and Paul Duncan, ed., *Michelangelo Antonioni: The Complete Films* (Cologne: Taschen, 2004), 53.

3

Imaginal Space and the Occult Soundtrack in Guy Maddin's *Keyhole* (2011)

Daniel Bishop

Since his emergence as an idiosyncratic arthouse *auteur* in the late 1980s, the work of filmmaker Guy Maddin has blurred the boundaries between autobiography and decadent fantasy, explored expressionistic continuities between his (frequently repressed or amnesiac) characters' internal and external landscapes, and surreally reclaimed archaic film genres and styles as objects of warped fascination.[1] While the occult imagination has always lurked in Maddin's films, his more recent works—in particular, *Keyhole* (2011), *The Forbidden Room* (2015), and the online *Séances* project (2015)—have pointedly foregrounded issues of the spectral. In *Keyhole*, ghostly domestic and psychic spaces overlap with recollections of *films noir*, "old dark house" movies, and Homer's *Odyssey*, telling the story of Ulysses Pick (Jason Patric), an amnesiac gangster and deadbeat dad who returns to his abandoned, ghost-infested home determined to reconcile with his estranged wife, Hyacinth (Isabella Rossellini).

The film opens in a barrage of gunfire, as Ulysses's gang fights their way past the police and into the Pick house. They lug with them a bound and gagged hostage: Ulysses's son, Manners (David Wontner), who initially goes unrecognized due to Ulysses's amnesia. Ulysses himself soon arrives with Denny, a blind and waterlogged young woman who refers to having just been drowned. We gradually discover that Manners and Denny (Brooke Palsson) are (or were?) lovers, and that Ulysses's plan is to use Denny's Tiresias-like clairvoyance to lead him through the house, recovering his lost memories by physically engaging with the space and its spectral occupants. In the meantime, Ulysses's gang while away their time in various pursuits: ghostly sexual escapades, an interior design project (led by an archetypically fussy, wallpaper-obsessed gang member), and eventually, an unsuccessful mutiny in which they attempt to kill Ulysses with a bicycle-powered electric chair. These narrative strands are, to some degree, tied together by obliquely expository voice-over delivered by Calypso (Louis Negin), Ulysses's wily, perverse (habitually naked) yet strangely sorrowful father-in-law, an older man who is chained to his daughter's bed, but seemingly able to transcend these limitations and cause mischief in his spectral form. As Ulysses works his way back to Hyacinth, a sense of their past life gradually emerges—an elusive and never entirely comprehensible picture, marred by familial conflict, sexual betrayal, and death.

In a 2012 interview, Maddin reflected on the spatialized nature of the film's experiential texture as intentionally blurring the boundaries between our feelings and its feelings, describing *Keyhole* as "a movie about a living space, about the emotions of space ... I wanted to really make a movie about our living space and the way we all feel about certain rooms."[2] In the final sentence of this quote, however, Maddin shows how easy it is to slip between describing "the emotions of space" and describing our emotions *about* space. Emotions, in this understanding, extend outwards from us spatially, not just a subjective experience but an objective manifestation that both is us and is other than us, a kind of reverberant echo of our presence that is shaped by (but also reciprocally shapes) our experience of the space it inhabits, perhaps even in the absence of the being who initiated it.

In this way, *Keyhole* engages something akin to philosopher Gaston Bachelard's phenomenology of the poetic imagination, particularly as it is explored in *The Poetics of Space*. Maddin has openly and frequently acknowledged the influence of this book on *Keyhole*, mentioning it in the film's press kit and in several interviews.[3] At the time of its original publication in 1958, Bachelard's interests had expanded from epistemology and the philosophy of science to encompass the ontology of the poetic image and, eventually, the phenomenology of imaginative experience itself. Taken as a whole, the book meditates upon the archetype of the house (and other spatialized domestic images, such as the attic, the drawer, or the corner), and explores their relationship with an imaginal state of being associated with intimacy and the physical act of dwelling.

Importantly, *The Poetics of Space* is not just a study of how spatial images act as literary tropes or symbols. Rather, it is a phenomenological exploration of the way in which, through these images, we experience the wholeness of our being as imaginatively interwoven into the world it inhabits.[4] These primal "direct images," which accommodate but essentially exceed definition by representational, ideological, or psychoanalytic meaning, become tangible through imaginative poesis, in both artistic representations that draw upon them, and the co-creative aesthetic bond between author and reader, a process Bachelard described with the aural metaphor of "reverberation."[5] The image of reverberation allows us to understand film sound as a model for how intimate domestic spaces are *felt*, for how these images are experienced in what might be called the poetic register of our being, and for how we poetically convey such images, thus intersubjectively instantiating experience in the being of another. In *Keyhole*, Bachelardian reverberation might also allow a more spectral, haunted reading, in which the spatialization of emotion and the spatialization of sound are interwoven. If, as Bachelard suggests, "the old house, for those who know how to listen, is a sort of geometry of echoes," then *Keyhole*'s soundtrack might be understood as manifesting the ghostly vectors that extend between and weave together the interior and exterior landscapes of its characters, simultaneously drawing its audience into a poetic apprehension of its haunted aesthetic world.[6]

Composer Jason Staczek, echoing Maddin, noted a desire to centralize the house in *Keyhole* such that music suggests the manifestation of autonomous (if not necessarily human) emotional agency:

> Guy and I actually discussed beforehand that we wanted the score to sound not like a musical score that had been placed on top of the film, but we wanted it to sound like it was emanating from the house. So that's what I was thinking about the whole time—if the house was making these groaning, broken noises and the sound was oozing out of the walls, and it was part of the ambiance instead of being there to comment on the story.[7]

Staczek's description captures the elusive quality of his score, in which musical elements as they are most conventionally understood—discrete pitches, rhythmic and melodic organization—act as modalities within a larger flow, materializing and dematerializing into a haze of oozing, clanking, murmuring ambience.

Film scholar K. J. Donnelly, writing about David Lynch's *Eraserhead* (1977)— another film of murkily ambiguous interiors—has offered the useful notion of a "sound continuum" in which music and sound effects are understood as poles of a conceptual spectrum, the middle space of which can generate and manipulate states of uncanniness and psychological discomfort.[8] To be clear: blurry boundaries between sound design and music, as Daniela Kulezic-Wilson and others have suggested, represent a much broader and more deeply rooted characteristic of contemporary cinema, extending across multiple subjects, genres, and styles that do not necessarily convey such particular effects.[9] But what distinguishes *Keyhole* within this broader aesthetic-industrial trend is its emphasis upon moments of acoustic materialization and dematerialization—of movement within Donnelly's spectrum as an index of spectral presence, thus using the soundtrack to manipulate the film's ambiguous boundaries between diegetic spaces and psychogeographic landscapes, between past and present, and between the living and the dead. This aesthetic is both a defining quality of the film's imaginal-physical space and a dynamic form of engagement mediating between the occupants of that space: the domestic objects it contains and the ghosts who seek engagement (whether compulsive or habitual, feverish or distracted) with these objects.

Keyhole's soundtrack thus serves as something akin to "acoustic ectoplasm," a fluid and mutable extrusion of visualized space that mediates between states of interiority and exteriority. This ectoplasm is distinct from (although certainly conceptually resonant with) the specific denotations of this term in spiritualism and parapsychology, in which it is understood as psychical energy manifesting in physical form. Film scholar Beth Carroll, in her recent study of the acoustic dimension of cinematic haunted houses, has explored a similar image, viewing sound as an ectoplasmic extrusion of the past, a ghostly continuance, distinct from its physical source, that reaches out to permeate the present.[10] Carroll's reading, however, depends upon a boundary, albeit permeable, between "the haunted" and "that which haunts," between an implied "us" and an implied "it." In this way, her discussion of "entrainment," of a sonic attunement shared by diegetic characters and filmic spectators, draws upon sound's ability to permeate our body as an ontological threat (or even as hostile, possessing force), manifesting in the affective charge characterizing the horror genre. *Keyhole*, by contrast, for all its uncanny qualities and its borrowing from the imagery of the haunted house film, is—simply put—not a horror film, with no interest in creating for its viewers an affective state of fear.

Nevertheless, the ectoplasmic permeability of physio-temporal spaces, characters, and objects is one of *Keyhole*'s most consistent and identifiable fixations, and not just on its soundtrack. Grasping at his lost memories, Ulysses inserts himself into a richly tactile sensory world, fumbling about in drawers, caressing wallpaper, huffing the fumes of long-abandoned catalogues, and habitually fidgeting with the hands of virtually every clock he sees.[11] Permeability thus functions less as a threat, and more as an ectoplasmic extension of an amnesiac characters movement toward the recovery of a forgotten domestic harmony, while ultimately acknowledging that the only way to recover what's been lost is, perhaps, to become a ghost oneself. This form of entrainment might be described less as a threatening "penetration," as in Carroll's reading and more as a Bachelardian "reverberation." *Keyhole* thus also fits within a contemporary swerve in millennial ghost stories, in which, as film scholar Murray Leeder has put it, "the line between our ghosts and ourselves is increasingly untenable."[12]

In the absence of theorization by Bachelard specific to cinema, we might also draw from more recent directions in film philosophy, to consider the hypothetical agency of "the film itself," as well as that of its diegetic haunted house. *Keyhole* is an excellent example of what film theorist Daniel Frampton has termed the "fluid filmind."[13] Frampton offers a particularly useful framework for understanding films whose spectral landscapes complicate our perceptions of "inside" and "outside." Within a "fluid filmind," whose projected reality is highly subjective and prone to sudden mutation, a haunted house might easily be regarded as a kind of diegetic avatar for the mind of the film itself. The house is thus a physical structure inside the macrocosmic film world, while also *itself* being a microcosmic world within which diegetic reality is characteristically fluid. Like a set of nesting dolls, the house, a microcosm of the filmind, is also a macrocosm to the microcosmic characters who inhabit it. The film-house is thus an animate, physicalized manifestation of the collective memories and emotions of its many ghosts. Extending this reading farther, we might also imagine these ghosts possess in turn their own internal landscapes, memories that might easily be projected outwards into diegetic space. Throughout the film, characters ambiguously populate each other's recollections, staging reenactments of seemingly past events that overlap with the film's (somewhat) more linear narrative present.

This ectoplasmic blurring is tangible within Staczek's score, as well as in the spaces between that score and its realization in the release version of the film, where it was extensively edited and supplemented with additional music by editor John Gurdebeke.[14] On one analytical level, we can identify these sonic qualities *within* Staczek's more traditional melodic and thematic musical elements—as, for instance, in the melody titled "Shaft," which serves as the film's equivalent to main title and end credits music.[15] Listening to "Shaft" in its foregrounded manifestations in the film (as, for example, at [0:01:18] and [1:30:22]), we hear three distinct elements: muted, pendular off-beats; a disjunct, leaping melody; and a murmuring, interspersed countermelody that coils repeatedly around a minor triad. Even these distinct gestures, however, are rendered sonically askew, with slight bends and blurs of pitch and rhythmic glitches. Staczek described a process of applying digital filters and other effects that would create small, randomized fluctuations: "I'd just throw things on the timeline, and if it was gooey or not quite right, I would sometimes go ahead and not try to make it line up any better, pitch-wise or rhythmic-wise."[16] The result is a melody that itself seems to shift in and out of focus.

This melody, however, is frequently used as a point of materialization within a far more nebulous ambient texture. This sometimes occurs as a result of Gurdebeke's post-production intervention, where he would drop fragments of Staczek's material into the soundtrack to occasionally thicken the ambient mix of sounds. In one key scene, however, the presence of "Shaft" fragments is distinctly the result of Staczek's own compositional agency.[17] When Hyacinth allows Ulysses to unlock the first of several doors, he enters an empty sitting room with Denny, dragging a still bound-and-gagged Manners (0:31:30). Musically, we hear a cue titled "Odds and Ends" ("Dissident et fines") on the soundtrack album. The scene plays out mostly wordlessly, through exchanges of looks, tactile gestures, and other sensory interactions between the human/ghostly characters and the domestic objects in their environment … a pile of catalogues (which Ulysses proceeds to sniff, as shown in Figure 3.1), a shoeshine kit, a taxidermized wolverine, the wooden dividers in a drawer, the worn nap on the arm of a sofa (which Ulysses scratches), and so forth.

Figure 3.1 (0:32:04) from *Keyhole* (Monterey Media, 2012).

The texture of Staczek's cue is sonically thick, but far more ambient than lyrically melodic. Sustained drones, sometimes brighter, sometimes more muted, overlap with distorted groaning sounds and translucent bell-like chiming. Other sound elements augment the texture: the uncanny tuning of the Manners's old radio, broadcasting a childlike voice narrating a story about a dog, and unintelligible, whispered fragments of a ghostly conversation between Hyacinth and her rebellious daughter Lota. Materializing and dematerializing within this haze of ambient elements, we also occasionally hear distorted fragments of the recognizable coiling countermelody from "Shaft." Repetitive, practically musicalized, the sound of Denny's furtive whisper ("Let's go. Let's go …") leads Ulysses out of the room.

These manifestations of recognizable melody do not "mickey-mouse" the characters' gestures toward embodied recollection. Other than one moment in which Hyacinth and Lota, suddenly aware that they are being watched, fade from sight accompanied by the cue's sustained dynamic climax, these moments of lyricism broadly and asynchronously parallel the action rather than capturing its moment-to-moment flow. In this scene, music (and musicalized voices) manifest against the ambient background like figures coming into focus through a field of fog. If we consider this manifestation through the metaphor of ectoplasm, we might understand it to also manifest spatially; not literally, in the sense of stereophonic manipulation, but rather in its use of sound to convey a sense of dreamlike, or imaginal space. Recognizable music seems to manifest *from elsewhere*, suggesting a larger concealment in a place of origin beyond the threshold of normal perception, materializing like the fragments of memories and physical sensations with which Ulysses, and with him our own filmic perception, become entangled, absorbed into the uncanny strangeness of banal, everyday objects and spaces of home.[18]

What Bachelard's imaginal houses and haunted houses have in common is that they are places of inversions, of present absences and darkness visible, a world in which the borders of inside and outside, object and subject, the house and the self, become dynamic and reversable. To put it in more occult terms: through the ghostly liminality of the soundtrack, the traditional "as above, so below" instead becomes "as within, so without." This re-phrasing is conducive to Maddin's sensibility, in which things lost (be they archaic film styles, repressed eros, or familial trauma) are always hovering on the sweaty, pulsating borders of melodramatic exteriorization. But they are equally conducive to Bachelard's reveries of domestic space and our co-creative entanglement with the reverberant images of home. This overlap with Bachelard's *Poetics* isn't without its tensions, but these tensions are productive, revealing something of the specificity of Maddin's aesthetic world. Bachelard's idea of "topophilia," his engagement with "felicitous space"—the spaces whose primal images of intimacy weaves them into the fabric of our imaginal being—is complicated by *Keyhole*'s emphases on trauma, regret, shame, loss, and corporeal embarrassments.[19] Maddin's career-long fixation on articulating these themes with frank, even squirm-inducingly autobiographical gestures is well-documented.[20] But so is Maddin's tendency toward unironic emotional sincerity that belies easy stylistic labels of pastiche or camp. If Maddin's film worlds thus ultimately preserve a kind of modernist authenticity by undercutting Bachelard's dreamy topophilia with a morass of sexual hang-ups and familial dysfunction, I believe that they offer us as consolation the ghosts of happiness—a consolation whose affective truth is self-consciously purchased at the cost of its ultimate insufficiency. "I'm only a ghost," says Ulysses, "but a ghost isn't nothing" (1:15:35).

As the film draws to its conclusion, it is ambiguously hinted that *all* the characters are ghosts of one kind or another. The gangster sub-plot fades from importance and the gangsters themselves gradually disappear, secreting themselves into wardrobes and the like. A phone rings. Manners answers, and the unheard message transforms his face into a rictus of despair. The soundtrack moves from the unstable, microtonal shifting of Gurdebeke's slowed-down "Glorious Cut 2," to the relative structural coherence of "Shaft," with its offbeat pulsations and its arching minor-sixth leaps, providing a sense of formalized return as Ulysses moves his things back into Hyacinth's bedroom, defeating her Homeric "suitor" by murdering her mute butler and lover, Chang (1:18:30–1:20:20). Manners's final interaction with Denny also suggests a cyclical temporality of return, in which the (now seemingly sighted) young woman cheerfully heads off to the midnight swim where she will drown. As the soundscape dematerializes into a haze of Staczek's music—vocal fragments, soft percussive and piano hits, and a slowly wandering drone pitch—Manners runs upstairs and collapses on his childhood bed (1:20:40–1:21:40).

This infantilized retreat leads into a dreamlike sequence, seemingly a memory of domestic tranquility in which Manners proudly shows Ulysses his model train set. Dad offers gruff encouragement while Hyacinth beams from her sewing desk. Staczek's cue, transitioning into this scene, presents us with winding arpeggios on vibraphone, clearly evoking Bernard Herrmann's score from the *Twilight Zone* (1959–64) episode "The Lonely" (1959), used as temp music in earlier working cuts of the film (1:21:36). Under this thin web of moving lines, however, the music gradually develops in a very different affective direction: a more churning, propulsive rhythmic drive begins to manifest as the family decides to ritualistically set the displaced objects in the room back in order (1:22:29). Propulsive forward movement thus frames a compulsive act of restoration, of turning back time and recovering the past. Once this momentum coalesces, we are offered perhaps the warmest, most appealingly generous musical gesture in the whole film, a lyrically arching, descending melody in an unusually warm, string-like voicing that briefly absorbs the entire soundtrack (1:23:27). It's the musical equivalent of an inviting light, seen through the window of a dark house at night.

And yet this strongest materialization of the film's ectoplasmic musicality is balanced by a merciless return to the film's (atemporal, circular, ambiently ambiguous) "present." Ulysses and Hyacinth, arm in arm, descend the stairs and tuck sleeping Manners into his bed. Numerous close-up shots (as seen in Figure 3.2) and Ulysses's never-ending fidgeting absorb us into this cluttered realm of child-like, *heimlich* bric-a-brac.

Manners then awakes to find the house resetting itself—gangsters vanishing, bullet holes fading into the woodwork. Staczek's score dissolves entirely, leaving the soundtrack entirely to Gurdebeke's "Glorious Cut #7," a roughly six-second repeating loop of an echoing *clang*, alternating with a series of dry metallic hits over a *hissing* sound, like a needle caught in the groove of an old record (1:27:10).[21]

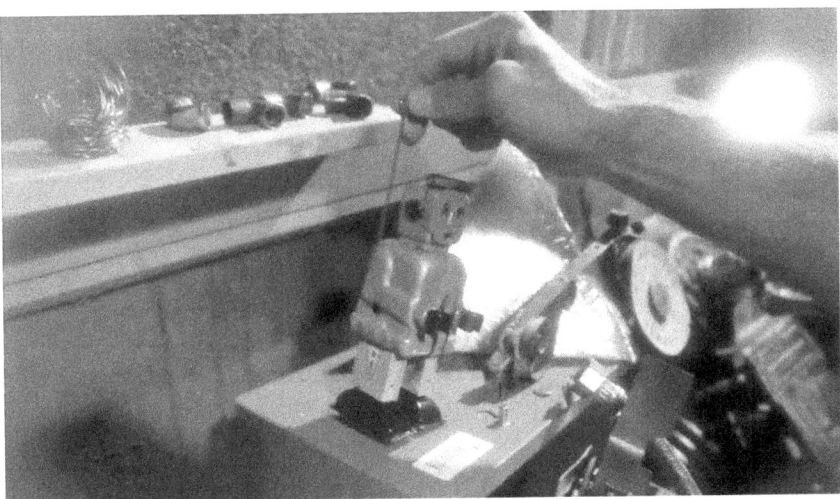

Figure 3.2 (1:25:28) from *Keyhole* (Monterey Media, 2012).

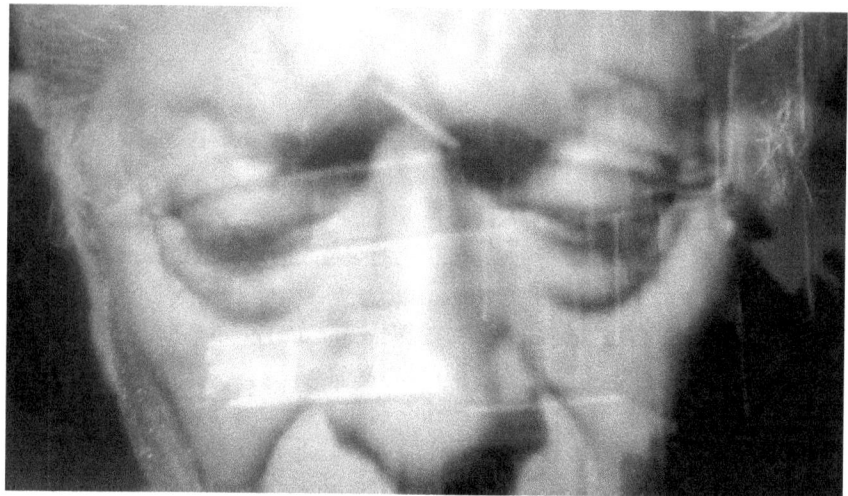

Figure 3.3 (2:08:34) from *To Kill a Mockingbird* (Universal Home Video, 2006).[22]

Framed by the disturbing and suggestive audiovisual fragmentation that bookends it, Manners's dream of domestic restoration conveys a strangely compelling moment of everyday transcendence, which the film takes utterly seriously. In an essay titled "Being Well-Lost in Film," film scholar George Toles (a frequent Maddin collaborator, and co-author of *Keyhole*'s screenplay) evokes a comparable image to explore the role of space within the imaginative dynamics of cinema. Toles's essay concludes with a poignant interpretation of the final sequence in *To Kill a Mockingbird* (1962). For Toles, Atticus Finch's house (see Figure 3.3) is an environment in which "we find an almost total harmony of the visible and moral realms, and both manifest the contained splendor of a quiet, well-lighted room viewed by a homesick stranger through a window late at night."

This stranger is, of course, local bogeyman Boo Radley, the counterbalance that, for Toles, allows Atticus's house its meaning and its potential truth.[23] Toles continues:

> Neither Atticus's goodness nor his profound commitment to justice is able to save Boo from the pain of his life, to relocate him by narrative slight-of-hand to the open space which is Atticus's home and identity. Boo is led back to the darkness of the Radley house by everything that is inside him … He is taken away from us, turned into the 'heart' of absence, so that Atticus's continuing *presence* can have a meaning that does not make false claims about the world.

In the imaginal space of cinema, Toles concludes,

> … the hidden must remain hidden … in a place where images cannot (will not) take us, if the well-lighted spaces that remain are to offer a comfort that endures. What is open to us in film owes its light to a neighboring darkness we are still able to feel.[24]

Keyhole offers us an internalization, or perhaps an "occulting" of Toles's argument, one whose borders are drawn not by the moral geography of childhood in a small Alabama town, but rather within the interior mnemonic landscape of a single home and its spectral inhabitants. Inverting Toles's reading, the affective force of Maddin's haunted darkness might depend upon the reality of some other world of happiness, however temporary. Manners's final dream manifests in the film world to the extent that it takes over, it *becomes* the reality of the film, expressionistically manifesting on the soundtrack a focused moment of concentrated, and highly unexpected sentimental lyricism, as if Atticus's world of unconditional love and benevolent order were somehow buried deep within the broken heart of the Radley home. And yet, while this image persists in a striking superimposed shot near the film's conclusion (Figure 3.4) of a sorrowful Calypso seemingly internalizing the Pick house within himself, that house is once again dark.

The sustaining light of Manners's happy memory is not a gesture toward perennial truth and essential selfhood, as Gaston Bachelard or *To Kill a*

Figure 3.4 (1:29:52) from *Keyhole* (Monterey Media, 2012).

Mockingbird might imagine it to be. But even if this recollection of harmony is lost to Manners at the film's conclusion, belief in its affective reality means that it might still be recovered. In the web of spectral time, the house is always waiting.

In the following extraordinary passage from *The Poetics of Space*, Bachelard seems to anticipate this dimension of *Keyhole*'s image world:

> In Henri Bosco's novel, *Hyacinthe* which, together with another story, *Le jardin d'Hyacinthe* (Hyacinth's Garden), constitutes one of the most astounding psychological novels of our time, a lamp *is waiting* in the window, and through it, the house, too, is waiting. The lamp is the symbol of prolonged waiting. By means of the light in that far-off house, the house sees, keeps vigil, vigilantly waits. When I let myself drift into the intoxication of inverting daydreams and reality, that faraway house with its light becomes for me, before me, a house that is looking out—its turn now!—through the keyhole. Yes, there is someone in that house who is keeping watch, a man is working there while I dream away. He leads a dogged existence, whereas I am pursuing futile dreams. Through its light alone, the house becomes human, it sees like a man. It is an eye open to night.[25]

Although Bachelard did not design his ideas to accommodate audiovisual or musical analysis, the aural sensitivity of his thought leaves significant (and largely underdeveloped) potential space for the exploration of screen media. Indeed, the manifestation of the soundtrack not just as imagined, but as a tangibly present, audible phenomenon arguably offers an even greater set of possibilities for intersection between the actual and the imaginal. With its ectoplasmic mutability, its imaginal tactility, and its spectral implications, the soundtrack of *Keyhole* reverberates across and through these boundaries, opening a rich conceptual space from which we might imaginatively explore haunted film worlds and our experiences of them.

Notes

1. For a concise mid-career summary of Maddin's biography, work, and aesthetic pre-occupations, see David Church, "Bark Fish Appreciation: An Introduction," in *Playing with Memories: Essays on Guy Maddin*, ed. David Church (Manitoba: University of Manitoba Press, 2009), 1–17.
2. Sam Adams, "Guy Maddin Talks about *Keyhole* and 'the Haunted House of Cinema,'" *The A.V. Club*, April 6, 2012, https://www.avclub.com/guy-maddin-talks-about-keyhole-and-the-haunted-house-o-1798230790 (accessed December 7, 2022).

3   The press kit is no longer maintained online. Prior to the production of *Keyhole*, Maddin discussed *The Poetics of Space* in two interviews from 2009 (in *Bomb Magazine* and *Rain Taxi*). The *Hollywood Reporter* review of *Keyhole*, following its premiere, refers directly to the press kit's citation of Bachelard. In two interviews from 2012 (from *The Skinny* and *Exberliner*), Maddin further enthuses about Bachelard. For full citations of these interviews and reviews, see bibliography.
4   Maddin does not discuss these aspects of Bachelard's philosophical project in any of the interviews cited above. However, given Bachelard's emphasis on the act of reading as ideally constituting active poetic engagement (what Mary McAllester Jones has described as his directive to read "pen in hand"), it is entirely appropriate to this project that Maddin is responding to Bachelard primarily as an aesthetic stylist. On Bachelard and writerly reading, see Mary McAllester Jones, *Gaston Bachelard: Subversive Humanist* (Madison: University of Wisconsin Press, 1991), 174–5.
5   On reverberation and intersubjectivity, see Gaston Bachelard, *The Poetics of Space*, trans. Maria Jolas (Boston: Beacon Press, 1994 [1958]), xvi and xxii–xxiii.
6   Ibid., 60.
7   Jason Staczek, interview by author, January 18, 2018.
8   K. J. Donnelly, *Occult Aesthetics: Synchronization in Sound Film* (Oxford: Oxford University Press, 2014), 131.
9   For a recent summary of these issues, see Daniela Kulezic-Wilson, *Sound Design Is the New Score: Theory, Aesthetics, and Erotics of the Integrated Soundtrack* (Oxford: Oxford University Press, 2020), 11–22.
10 Beth Carroll, "Acoustic Ectoplasm and the Loss of Home," in *Film and Domestic Space: Architectures, Representations, Dispositif*, ed. Stefano Baschiera and Miriam de Rosa (Edinburgh: Edinburgh University Press, 2020), 106–17. Darren Werschler has also explored ectoplasm as an image in Maddin's work. See *Guy Maddin's "My Winnipeg,"* Canadian Cinema #6 (Toronto: University of Toronto Press, 2010), 110–15.
11 In this way, *Keyhole* might also engage recent critical perspectives on haptic cinema and phenomenology. On the haptic soundtrack, see Kulezic-Wilson, *Sound Design Is the New Score*, 36–7. Recent work by film scholar Saige Walton has explored a related dimension of this tactile exchange by applying the phenomenology of Maurice Merleau-Ponty to explore gestural, or baroque cinema, as well as the atmospherics of mood. Walton's body of work, at varying points, has addressed Guy Maddin and Gaston Bachelard. See Saige Walton, "Hit with a Wrecking Ball, Tickled with a Feather: Gesture, Deixis, and the Baroque Cinema of Guy Maddin," in *Playing with Memories*, 203–23. See bibliography for other full citations.
12 Murray Leeder, "Contemporary Ghosts," in *Twenty-First Century Gothic: An Edinburgh Companion*, ed. Maisha Wester and Xavier Aldana Reyes (Edinburgh: Edinburgh University Press, 2019), 139–40.

13 Daniel Frampton, *Filmosophy* (London: Wallflower Press, 2006), 88–90. Frampton's neologism, combining the words "film and mind," is a hypothetical entity that is both "the film itself," and an agential *presence* whose characteristically intentional mode of self-perception underlies filmic representation.

14 All references to the film with timings below are citations of the film's home video release on DVD, in (hour: minute: second) format. *Keyhole* (2011), dir. Guy Maddin, DVD (Monterey Media, 2012). I am deeply thankful to Jason Staczek for providing comparative access to a pre-release cut of the film, as well as a production cue sheet to inform my detangling of this soundtrack's complex musical authorship.

15 This track, independent from its motivic manipulation elsewhere in the film, was part of a package of pre-existing music provided as potential temp music from earlier projects, although it remained in the final cut of the film. It is thus appears in *Keyhole*'s music credits by this earlier title, independent from the rest of the score. Staczek retitled the cues into Latin (sometimes loosely—"Shaft," for example, becomes "Saggita") for the soundtrack album. See *Music from the Motion Picture "Keyhole,"* CD (Milan Entertainment, 2012).

16 Jason Staczek, interview by author, February 11, 2022.

17 To be clear, this wording is not intended to have qualitative connotations that inherently privilege less-modified examples of Staczek's scoring. In fact, discussing this aspect of the soundtrack's authorship with me (while looking at clips that he hadn't seen in years), Jason in fact seemed open-minded and genuinely interested in how his unique work had manifested, in some cases beyond his direct creative control.

18 Assistant editor Mike Olenick has described this scene as particularly challenging to his task of preparing a coherent rough cut, due to the lack of scripted dialogue. Needing to dig deeper into the rushes to provide the scene with coherence led Olenick to go beyond the scripted version in a way that arguably also enhanced its focus on the characters' sensory interaction with their environment. See Mike Olenick, "Cutting, Connecting, and Conjuring: Editing with Guy Maddin," *Wexner Center for the Arts,* 18 June, 2020, https://wexarts.org/read-watch-listen/cutting-connecting-and-conjuring-editing-guy-maddin.

19 Bachelard, *The Poetics of Space*, xxxv–xxxv.

20 See Donald Masterson, "My Brother's Keeper: Fraternal Relations in the Films of Guy Maddin and George Toles," in *Playing with Memories*, 26–47. See also Werschler, *Guy Maddin's "My Winnipeg,"* 16–36.

21 "Glorious Cut #7" is also slowed to 20 percent, a technique used with several of Gurdebeke's additional music selections. This manipulation of playback speed further foregrounds the tracks' microrhythmic and microtonal granularity to uncanny effect.

22 *To Kill a Mockingbird*, dir. Robert Mulligan. DVD (Universal Home Video, 2006).
23 Although these two films manifest very different musical worlds, both engage significantly with the themes of memory, family, and domestic space. For a very different, yet highly compelling reading of *To Kill a Mockingbird* along these lines, see Berthold Hoeckner, *Film, Music, Memory* (Chicago: University of Chicago Press, 2019), 201–29, especially 226–9.
24 George Toles, "Being Well-Lost in Film," in *A House Made of Light: Essays on the Art of Film* (Detroit: Wayne State University Press, 2001), 47–8.
25 Bachelard, *Poetics of Space*, 34–5. This passage also demonstrates Maddin and Toles's (perhaps sly, perhaps coincidental) appropriation of specific images and even the name of a major character from Bachelard's text.

4

The Haunted and the Medium

K. J. Donnelly

The VVitch (2015), subtitled "A New England Folktale," takes a significant interest in landscape. It appears to provide an alternative foundation myth for the United States of America. In 1603, austere Protestant William (Ralph Ineson) and his family are banished from a settlement in New England. Alongside his wife and five children, they build a homestead in a remote area next to a dense, dark forest. A succession of strange events befalls them: their crop fails, their baby disappears and one son, Caleb (Harvey Scrimshaw), appears to be possessed by a spirit after visiting a witch's house in the forest. The young twins Mercy (Ellie Grainger) and Jonas (Lucas Dawson) accuse their older sister Thomasin (Anya Taylor-Joy) of causing the events through being a witch herself, and the parents also begin to believe this. Death befalls all of them except Thomasin, who, at the conclusion appears to join a coven of witches in the forest.

While the film is rich in thematic ideas and open to many interpretations, it retains a high level of ambiguity about events depicted, which become even more uncertain with repeat viewings, although what becomes clearer is the importance of the film's soundtrack. *The VVitch* leans heavily on location and sound. Representational presences and absences are personified sometimes directly by the music. The film's music not only embodies the wooded location and manifests the threat of evil in the film but also stands in as an uncertain presence for not only the witch but also for the anxieties of the family. Most of the music consists of eerie drones, with the timbre dominated by wooden bowed string instruments, where we can hear the wood and the metallic scrapings of the strings, making for a very human and arboreal presence, and manifesting a threat and perpetual anxiety. Audiovisual landscapes regularly include a significant sonic component, which can supply an important uncanny aspect to them, particularly as sound is considered more ephemeral, and less clear and concrete than the visual. As "background," sonic and visual landscapes can contain a potentially determining psychological charge that can dominate proceedings or at cast a shadow over the rest of the drama, as happens in *The VVitch*.

Visual Absence and Sonic Presence

In *The VVitch*, at certain points a breakdown takes place, pulling apart sound and image. This is the collapse of the "audiovisual consensus" of sound and image matching and cohered into a seamless unity, and this uncoupling suggests an "abnormality" and sense of psychological disturbance without laboring the point. The film inaugurates with the lone acousmatic voice of William, addressing the town elders. It is accompanied by a sustained frontal image of his daughter Thomasin, succeeded by shorter shots of her brother Caleb and then the twins Mercy and Jonas in their mother's arms but failing to show her face. This is a succession of reactions to William's continuous monologue. While we see the family's faces, they are succeeded by a curious straight shot of the back of William's head. It seems like he is on trial and yet he is berating and questioning his inquisitors. When the village elders inform him of the family's banishment, we have a 90-degree camera angle of Thomasin in the foreground and William in the background replying that he is happy about the banishment. When the camera finally alights on William's face he is no longer speaking. This is extraordinary in aesthetic terms. This sequence may tell us about the event setting the narrative into action, but it is at least as significant in its care with, and significance assigned to, audiovisual aesthetics. It tells us how important William's voice is, but it also shows us sound uncoupled from image.

This, in notable fashion, "mirrors" the film's penultimate scene, when Thomasin invokes Black Philip (Wahab Chaudhry) and we hear his acousmatic voice, until finally seeing him obscured in semi-darkness. This comes as a surprise as we hear his voice yet retain the shot of Thomasin to the point where we wonder if he is in her imagination or not. When he finally, and perhaps unexpectedly, appears on screen he is obscured in a similar fashion to that of the first appearance of the witch. We never see his face very clearly, but his thin facial features resemble William somewhat in the darkness. Indeed, the film has something of a palindromic form, not only in this aesthetic strategy but in terms of both situations embodying the leaving of one "culture" for another as significant transitions: leaving "civilization" for a lone homestead, and foregoing that for a life as a witch.

Both sequences evince a "pulling apart" of sound and image tracks, suggesting a discontinuity rather than simple cohesion, and problematizing the illusion of a coherent diegetic world. Indeed, one striking aspect of *The VVitch* is how little is seen of her. The film instead projects anxiety onto the landscape, through music as an important agent of this process. Events might be explained away as fantasies and dreams, yet the music provides a constant reminder of the threat and anxiety, halting any thought of dismissing proceedings and being a material form of the film's dread in itself. Gaps produce structural absence, which helps add up to a form of presence. In *Making Meaning*, David Bordwell's discusses four levels of understanding used in film comprehension and analysis.[1] The final one, symptomatic meaning, is where analysis assumes that the film is hiding something; he states that this is not a valid form of film analysis. However, there are many films that neglect to show and tell everything, and some that like to extend themselves through resonance and vague reference to absent things. Subtext can sometimes be intermittent and incomplete but insistent. The audience also mentally fills in what the film does not make explicit, and this is clearly the case with *The VVitch*, where there are many gaps and ambiguities. Furthermore, sound and music can have an ambiguous presence and can transcend a limited function, hinting at memories and making half-implications that may or may not be picked up by the audience.

Absence or partial absence can nevertheless have an effect, while there can be an uncertain effect from something that might seem insubstantial. Incidental music in film is often considered as "background" or "window dressing," yet can have a powerful effect, often unconsciously. Hauntology has developed as a concept since Jacques Derrida first proposed it in *Specters of Marx*.[2] At heart, it retains something of Derrida's more general concerns with the uncertainty of communication, though. Mark Fisher, Simon Reynolds, and Jamie Sexton have used it for a sense evident in music of aesthetic traces of the future that was proposed in the past but never materialized.[3] A wider understanding would register that ghostly activities represent an unresolved past while simultaneously being symptoms of repressed or forgotten knowledge. However, I would suggest that audiovisual culture has its own particular sense of hauntology, which derives from the material nature of sound and image technologies being yoked together, with the constant threat of dissolution of the illusion and collapse of the medium into chaos. Consequently, films and other audiovisual culture contain points of ambiguity that arise between the modes of sound and image and might fail to be confirmed through "cross-modal" reference.

Some films have soundtracks that can adopt a position of relative autonomy in relation to the image. This was always a capability, evident in although rarely exploited by the structural position of the non-diegetic film score in Classical Hollywood films and before that, in live music produced *in situ* when a silent film was screened. Sound and music are able to imply or suggest something that is not present or not connected to the images on screen or dialogue, the primary zone of anchored meaning in film. This does not have to be "information," it can be a feeling or more distinct emotion, or a sense of unease, and perhaps due to its lack of corroboration in the image and dialogue, this can become powerful and exert a significant cast upon the proceedings. As such, filmmakers are careful to avoid these if they can. As I have suggested in *The Spectre of Sound*, non-diegetic music can constitute a coherent discourse in films and:

> … might be conceived as the virtual space of mental processes, making film music the unconscious space of the film. It may be more like an unconscious level of the film (or a level of unconscious in the film). We can then see it as a repository of reminders, half-memories and outbursts of emotion and the illogical. These "ghosts" and "memories" that can haunt a film are often little more than shapes, momentary configurations or half-remembered sounds.[4]

So, this is a different form of hauntology from that which provides a reciprocal presence through its absence, but more one that has an uncertain presence and ambiguous value in the equation of audiovisual culture. In *The VVitch*, music can manifest a sense of presence for the almost wholly absent negative force as we see little supernatural in the film.[5]

Presence in absence might be understood as central to hauntology. In sound and music mixing, there is a curious sonic phenomenon called "presence," whereby, through boosting mid-range frequencies a recording can appear to be closer to us than before. In *The VVitch*, the music has its own presence, boosted at times by a sense of composer Mark Korven playing the instruments alone, rather than an anonymous professional orchestra or digital technology. For the film, the music furnishes a ghostly presence, compounded by it appearing to substitute for the witch, whose presence is not substantial in representational terms, comprising inconsistent fragments and with different appearances. Presences and absences are embodied sometimes directly by the music, where it is an agent for the malevolent presence, be that the isolated instances of the witch or the omnipresent threat from the forest.

Ghosts and the Nonindifferent Medium

The sense of integration between landscape and music (and other sound) is hardly new. Sergei Eisenstein was interested in their expressive possibilities, discussing what he called "nonindifferent nature" where landscape shots had an emotional function like music and lacking the burden of providing narrative information. Rather than being neutral and representational, such images carry a suitable emotional charge which is in tune with the rest of the film. He calls this "the musicality of landscape."[6] Eisenstein's notion of "nonindifferent nature" posits an equivalence between landscape shots and music.[7] Both are more than simple backdrop and supply an emotional charge and valence for the film. Both can move beyond simple communication and narrative function to elicit a sense of thick atmosphere and emotion. Eisenstein was interested in "emotional landscape shots" and their ability to step outside of narrative and diegesis. Later he was concerned with sound film beyond basic "talking films" and often thought and theorized in musical metaphors.[8] "Nonindifferent nature" posits an equivalence between landscape shots and music. Both have the capability of being more than simple backdrop to characters and action in the foreground and supply a significant emotional charge and valence for the film. Straightforward communication and narrative function are shelved in order to elicit a sense of thick atmosphere as well as emotion, in a startlingly similar manner to music.[9] These "emotional landscape shots" have an ability to step outside of the narrative focus and make further implications or suggestions.

Landscape shots matched with non-diegetic music could potentially redouble this effect. Rather than simply a representation of location, film landscapes with music can transform into something extraordinary. Through building their own integrity they might even "step outside" the film. Eisenstein appears to pull back from this, though, and focuses on landscape shots' place within the film system, in relation to other shots and elements, where they offer emotion to their surrounding pieces of celluloid. Yet, they can also be understood as something-in-themselves, as semi-autonomous objects within film. So, this is not about "representation" as such, and the music does not "add" to the landscape image, but rather converts it, perhaps into an "emotional representation," which is less of a representation of a place than it is a representation of an emotion. Even the most specific footage of a place can become notably different as film and music combine.

Eisenstein's suggestions, both about landscape shots and about the power of music in film, imply that there is something *in* the materiality and techniques of film that enables the production of something special.[10] Indeed, there is something in the medium itself. Since its inception, mechanical and electronic recording of sound and image has had a sense either of capturing or embodying a spectral world beyond the everyday. This characteristic has been identified in early film by Tom Ruffles in *Ghost Images: Cinema of the Afterlife*,[11] where he argues convincingly that the depiction of existence after death was a prominent theme in the early years of film. Similarly, uncanny aspects have been identified with the coupling together of synchronized moving image in early sound cinema by Robert Spadoni in *Uncanny Bodies*.[12] Tom Gunning noted the "uncanny powers of the cinema,"[13] while in *Haunted Media*, Jeffrey Sconce pointed to the consistent association of the uncanny and ghostly with new instances of electronic media and communication technology.[14] Indeed, there is a strong tradition of recognizing a deeply ingrained connection between audiovisual culture and the supernatural or paranormal. All these scholars point out the seemingly bewildering but compulsive character of the media, particularly the dawn of cinema and the transition to sound cinema beginning in the late 1920s. This makes sense in terms of new ways of seeing and hearing seemingly breaking the "natural laws" of our senses, as well as reordering time and space radically. One of the earlier theories associated particularly with the audiovisual was the "Ghostly Effect" suggested by Eisler and Adorno, who saw music as a way of charming away the disturbing fact of films showing ghosts of the past and animating specter-like shadows on a wall.[15] Yet in horror films, this might be constituted as a "reverse Ghostly Effect," where the sound and music, instead of calming and pacifying us, exploits our anxieties and amplifies them. Concurring with these theorists and historians, there appears to be an inherent potential in the medium of audiovisual culture for strangeness and the uncanny, if not possessing an uncanny character in itself.

Haunted Landscape in *The VVitch*

Building on Peter Hutchings's writing, Paul Newland notes that menacing landscapes on screen hold a "dark heritage" and are suffused with anxiety.[16] *The VVitch*'s setting is of paramount importance. It is a seemingly remote and bucolic setting: a "clearing by the woods." This embodies a geography that pervades folk tales and folk horror films, that of the dual location of "cultivated civilization" neighboring the "wild wood," with its dangers both known and unknown. One is open and light, the other dark and obscured. It opposes the wood and the witch, with the homestead and the austere Christian belief of the settlers. Initially, this seems a traditional tale about the wilderness and attempts to tame it. The landscape in *The VVitch* is actively uncanny, encompassing the soundtrack as an integral part of it, and in neither case forming a simple, passive backdrop to action but being an active part of the drama.

Geographer James Duncan asserts that landscape acts as a "… signifying system through which a social system is communicated, reproduced, experienced, and explored."[17] Therefore, it should not be considered inert, or simply a backdrop, or reflective, but an active agent. Furthermore, particularly for Europeans, the forest holds an ancient importance. In films, it also has a valence. For Europeans, the dark forest has been a perennial source of folkloric anxiety as well as a site of imagination,[18] and in films it retains something of the folk horror character of embodying an unknown threat from the past to the present.[19] The audiovisual forest is a particular, spatialized, psychological manifestation, established through audiovisual style. By six minutes into the film, we twice have long shots of the forest accompanied by ominous music. Stylistically, the film constantly uses very, very slow zooms (which in many cases were digital achieved in post-production) and slow dolly zooms. This gives an uncanny feeling to the location.

One of the most striking shots in the film is of the family on their knees, praying in the direction of the forest. They are small next to the landscape and are framed from behind showing the aim of their prayers. Are they praying to God or to the forest? William's particular brand of Christianity is enough for them to be removed from the settlement, perhaps it is even heretical. Consequently, these woods are given a special, supernatural status early on, one that is compounded by the film's creation of a sense of materiality and of agency to them. This striking shot is echoed later when the dog Fowler looks at the hare in the forest. In this first shot, as the family pray toward the forest, the camera zooms above them at the forest itself, and we hear the female choir music that is associated with the witch. This illustrates clearly the way that the music materializes a presence where there is an absence.

A startling sequence is where Tomasin plays "boo!" with baby Samuel, and indeed this featured in trailers and publicity. In a succession of reverse shots with her straight on to the camera, the final revelation of what she sees is not

Samuel but an empty space where he should be. The next shot shows her running toward the woods, even though we have seen nothing take the child or move in that direction. Thomasin halts before the woods subsequent to a long shot in which a figure in a red riding hood can be seen carrying the baby through the dense woods. This constitutes more than a single shot and is not from Tomasin's point of view, which suggests it might be a fantasy. The following shots show the baby from a flat angle and an obscured view of the witch covering herself and her broom handle with blood, accompanied by choir singing and a wooden beating sound.

This sequence begs a question. Is the witch a person? Or imagination? While the witch is not signaled as Native American, she is "native" in the way that she is already there. Prior to this, however, the film has carefully shown us Native Americans entering the settlement stockade, looking back ruefully at the leaving family. Perhaps the witch makes more sense as an organic force of the forest. In ancient Roman mythology, a minor deity/spirit who protected a specific place was called a *genius loci*.[20] Thus, the witch might be understood less as a person and more as an embodiment of the forest location, and so literally "organic."

This Wilderness … Will Not Consume Us

Forests have a specific sound. This usually involves the deadening of close sounds, where the lack of reflective surfaces added to the sound-dampening effect of trees and leaves, supplies what sound recordists call a "dead sound" with little sense of space or dynamics. Added to this, forests can also give a sense of space through distant sounds, where sounds in clearings and at the edge can be amplified and sound louder when you are inside. In *The VVitch*, the woods have very little in the way of sound. Instead, a soundtrack is provided by the non-diegetic music. The musical score for the film is dominated by wooden instruments, which yield a quite "wooden" sounds, particularly as the instruments are recorded to emphasize their materiality (scraping, key noise, etc). Therefore, in a way the music is a constant reminder, or perhaps even material presence of the forest throughout the film. The forest may be dark, with visibility impaired and sounds distant, but nevertheless they are enveloping and consuming both visually and sonically.

For a film about a witch, we are given surprisingly little in the way of one on screen. Indeed, only a handful of short and sometimes obscured sequences, and the also obscure images of floating witches at the film's conclusion. This amounts to a structured absence which is a defining aspect of *The VVitch* and suggests something of the film's central concern, about the accepted known world and the unknown. In the regular absence of the witch on screen, the soundtrack creates a material manifestation of her threatening presence. This is not an uncommon strategy in horror films, particularly for ghosts. Certainly, the film is so ambiguous about her nature that she could even be construed as a ghost. Thus, music has a clear importance for *The VVitch*.

Instead of a "regular horror score" the film includes much in the way of disturbing sounds, at times blurring any distinction between non-diegetic music and diegetic sound effects. The score has novel functions (not as continuity, cohesion, themes, emotional, and excitement effects) but as direct psychological state "imposed." It is overwhelmingly emotional, although at times seeming to be representational. The uncanny effect comes from the mix of musical functions and diegetic sound functions, where expectations related to traditional function are reversed, reformulating the norm that diegetic sound functions as "representation" and incidental music as primarily "emotional."

As I noted earlier, despite being unconventional in most ways, *The VVitch* follows two traditional horror film strategies, that of siting a threat of evil nearby in the woods, and of manifesting that evil in the film using non-diegetic music. However, *The VVitch*'s highly effective musical score, by Mark Korven, is far from a traditional or conventional component of the film.[21] *The VVitch* eschews the use of a conventional musical score and indeed, its attitude to sound generally is unconventional. There is an absence of strange or disturbing supernatural diegetic sounds. There is also little in the way of acousmatic sounds (scary sounds with ambiguous origins). In terms of the score, there is little in terms of "making the action," which in horror films means tension and release structures built around drones and *ostinati* (looped musical figures) for tension and then a stinger, a blast of musical sound as a conclusion. At least partially, this is due to the film not relying on set piece sequences.

As a counterpart to the organic horror held within the dense evergreen forest, the music is premised upon organic, fundamentally acoustic sounds. The organic sources of the music marks a sense of the "authentic." Indeed, the instruments are audibly physical manipulated, making them more personal and human and contrasted with the more formal, impersonal sound of most film scores. Even at points in the film where we hear a regular beat, this was achieved through manually hitting an instrument rather than using drums or electronics.

One of the most striking things about the sound of *The VVitch* is the use of old instruments that appear to have been repurposed, been made to do something "unnatural." The score is far more interested in sonorities than melodies and aims at momentary and overall sonic texture rather than being in any way memorable. The dominant musical sound is that of bowed strings. Nonetheless, rather than the banks of orchestral violins we are used to hearing in films, the instruments are often played solo, and recorded in a manner than catches and emphasizes the sound of instrument resonance, as well as the extraneous sounds, such as the buzz on the wood when the strings are held to the instrument necks.

The sonic character of the film is defined by the use of particular acoustic instruments in Mark Korven's score. The range of instruments includes nyckelharpa (and some bass nyckelharpa). This is a keyed violin, where the metal strings are depressed by hard wood rather than soft human fingers. The four strings are bowed manually but it has a "woody" sound due to the strings contacting wood, and it also includes twelve resonant droning strings which retain the same note and provide extra resonance.[22] Korven notes, "The backbone of the score was a Swedish instrument called the nyckelharpa. It's a medieval keyed violin and when Rob [Eggers] first heard it he said, 'That's it, that's the sound of the score.'"[23] It is crucial to note here that the director was interested in a particular "sound" rather than using a readymade sound, such as the orchestra, and conceived of the film as a particular timbral environment. The nyckelharpa is an eccentric choice as it has no connection to the representations in the film. It is from the same period but was not connected to Britain or America but rather Scandinavia and Germany. Lacking historical or representational relevance, the sound is important in itself and as such it provides a highly distinctive tone and character for *The VVitch*.

Korven played most of the instruments himself, including a cello and a hurdy-gurdy.[24] The hurdy-gurdy is a bowed string instrument. It is not played with a hand bow, rather it is instead sounded by a hand cranked wheel, which rubs on the two principal catgut (or nylon) strings. The strings are shortened or lengthened by wooden blocks controlled by a small keyboard. Although extremely popular in the Middle Ages and Renaissance, it later fell into obscurity. Like the nyckelharpa, the sound of bowed strings is far "woodier" than contemporary strings and has four single-pitch drone strings which provide part of the characteristic buzzing drone from the instrument's bridge. These bowed stringed instruments provide a good deal of the characteristic tone of *The VVitch*, while Korven brought in a professional specialist to play the jouhikko, a Finnish horsehair-stringed bowed lyre from Finland, which is used for the witch and is perhaps most clearly heard when the baby that is stolen.

Much of the music has the quality of what might be construed as noise. This gives it the added sense of being a permanent ambience. The score uses other instruments such as the waterphone, which is a modern instrument based around bowing rods on a metal container containing some water (called a "friction ideophone"). The whole of the device vibrates, providing an ethereal sound, and being characterized by its particular type of resonance. Indeed, the score regularly includes sustained notes and resonances, while yielding a sense of stasis and texture.

The music is also "organic" in a sense, in that it was largely improvised and refuses to use repeated themes to support the narrative development but instead plays upon the production of particular sounds using antique instruments. It has something of an informal character, too, in that a lot of it could conceivably have been produced spontaneously by Korven alone as live accompaniment to the film's action.[25] The highly distinctive sounds are featured "in themselves" rather than as component parts of melody, harmony, or rhythm. The score consists of acoustic instruments and focusing on their distinctiveness and resonant capabilities.

The sense of an organic sound appears to be contrived. Much of the sounds emphasize the wood of the instruments, which we can hear as sound box resonators but perhaps more notably we can hear the vibration and rub of strings against wood. Arguably, this is clearest with the nyckelharpa. This sound produces a doubling of the wood in the film, providing a strong sense of the forest location. The sources of the music (predominantly old and antiquated instruments), therefore mark a sense of the "authentic" paralleling of the sounds of the acoustic vibration of wood, metal, and string as a homology or perhaps an emanation from the charged forest location on screen. As already noted, the instruments tend toward drones and resonances, some of them having dedicated drone strings. Acoustic space is highly evident in the recording, as the instruments were not purely close-miked in recording and so allowing a sense of space to them that is bolstered by electronic reverb in the final mix, too.

Voices are important in the film, although the family depicted is hardly a garrulous family. There is little dialogue, yet William's highly distinctive monotonous but sonorous baritone voice and Yorkshire accent appears masculine and authoritative. This is echoed in the sound of Black Philip's soft and sonorous voice at the film's conclusion. The twins sing songs in tribute to the power of Black Philip, which seems incongruous and hardly innocent, yet is not halted by their parents. Human voices also have a crucial structural function. A choir of female voices appears alongside the witch, whether accompanying visual appearances or implied presence. The singing has an avant-garde style based on the singers improvising to vague instructions.[26] With its clear but undulating and discordant tones, although human, it could perhaps be considered more "angelic" or supernatural in character. The swirling glissandi and echoed sustained vibratos are atonal rather than based on traditional musical harmony. The words were taken from an actual witches' rite. We first hear the singing over a shot of the forest, and later when the witch uses baby Samuel's blood to begin to fly, when Caleb encounters the witch at her house, and at the conclusion where Tomasin joins the coven. It sounds remarkably similar to the vocal sounds of György Ligeti's *Requiem* (1963–5) and *Lux Aeterna* (1966) (which has sometimes called "micropolyphony") as used in Kubrick's *2001: A Space Odyssey* (1968).[27] Like *2001*, *The VVitch* has an ambiguous and confusing ending, unless we take it at "face value," which ultimately is not easy.

Conclusion

The old adage that music in a film either can function as set or camera, in other words as an agent of narration or of atmosphere and location, is particularly apt in relation to *The VVitch*. Despite this, the music merges with landscape as audiovisual agent, and a dramatic agent in itself rather than being a part of narration's telling of the story. Following horror film tradition, the music depicts something malevolent attacking the family. It also embodies the family's emotional situation, which is a more general tradition for film music. Further to this, it is an agent for embodying the location. This is rarer in films. In *The VVitch*, however, these three are all in effect the same.

The film's soundscape has more far-reaching effects. It furnishes a continuum that makes us miss the ambiguities of the film's narration, while the music's relative autonomy aids the drifting apart of sound and image, where music at times can appear almost autonomous. Despite this divergence we still try to compute as a unity; we *try* to integrate the experience. *The VVitch* appears based on an idea of doubles[28] most evidently in its rendering of the "w" in the title and the young twins. The name "Thomasin," as a female version of Thomas, means double or twin. Are there two Thomasins? Or she is double of the witch? The witch appears in two versions (young and old), who are not necessarily the same person.

The VVitch is rich in thematic ideas and open to many interpretations. With themes of Empire and colonization, fundamentalist religion and the nature of female sexuality, the film allows, perhaps encourages, different interpretations. Nonetheless, for what is on the surface a horror film, we see very little that is supernatural. Instead, the film's soundtrack provides a material form of the paranormal and its associated anxieties, haunting the landscape and embodying the presence of the witch and a pervasive negativity toward the family. The witch herself is only shown a handful of times in the film, often obscured, and when she appears, there is something not quite real about her. The film is haunted by the specter of this witch, who is manifested perhaps most vividly in the soundtrack, and in the threat posed by the forest, all three of which form a coherent discourse of anxiety. Mark Korven's music has more of a spatial than temporal function. It manifests a continuum of place rather than marking progressive time in the film. It is more interested in the particularities of sounds and marking a miasmic presence than being "conventional music" and developing through adopting the functional character of film narration. It seemingly lacks a sense of musical structure. A certain sense of temporal sonic structure is provided by the images and narrative development, yet this is a weak link, while its strong sense of space is formed by merging with and embodying of the location of the film, as an "environmental" element that makes the surrounding landscape uncanny.

Notes

1   David Bordwell, *Making Meaning: Inference and Rhetoric in the Interpretation of Cinema* (Cambridge, MA.: Harvard University Press, 1991), 8–9.

2   Jacques Derrida, *Specters of Marx: The State of the Debt, The Work of Mourning and the New International* (New York: Routledge, 2006).

3   Mark Fisher, "What Is Hauntology?" *Film Quarterly* 66, no.1 (2012): 16–24; Mark Fisher, "The Metaphysics of Crackle: Afrofuturism and Hauntology," *Dancecult: Journal of Electronic Dance Music Culture* 5, no.2 (2013): 42–55; Simon Reynolds, "Haunted Audio, a/k/a Society of the Spectral: Ghost Box, Mordant Music and Hauntology," *The Wire*, no.273 (November 2006): 26–30; Jamie Sexton, "Weird

Britain in Exile: Ghost Box, Hauntology and Alternative Heritage," *Popular Music and Society* 35, no.4 (2012): 561–84.
4 K. J. Donnelly, *The Spectre of Sound: Film and Television Music* (London: BFI, 2005), 21.
5 Saige Walton states that "In *The VVitch*, the supernatural is portrayed as something that is emitted or, more precisely, exhaled by the film's environment." "Air, Atmosphere, Environment: Film Mood, Folk Horror and The VVitch" in Screening the Past, Issue 43, April 2018. http://www.screeningthepast.com/2018/02/air-atmosphere-environment-film-mood-folk-horror-and-the-vvitch/ (accessed February 20, 2022).
6 Sergei M. Eisenstein, *Nonindifferent Nature: Film and the Structure of Things*, trans. Herbert Marshall (Cambridge: Cambridge University Press, 1987), 389.
7 Ibid.
8 Ibid.
9 Later Eisenstein was concerned with sound film beyond basic "talking films" and consistently thought and theorized in musical metaphors.
10 Saige Walton notes that another film theorist, Béla Balázs, points to film landscapes as having a "non-human sensuality." Walton, "Air, Atmosphere, Environment".
11 Tom Ruffles, *Ghost Images: Cinema of the Afterlife* (Jefferson, NC.: McFarland, 2004).
12 Robert Spadoni, *Uncanny Bodies: The Coming of Sound Film and the Origins of the Horror Genre* (Berkeley, CA.: University of California Press, 2007).
13 Tom Gunning "To Scan a Ghost: The Ontology of Mediated Vision," *Grey Room*, no.26 (Winter 2007): 96.
14 Jeffrey Sconce, *Haunted Media: Electronic Presence from Telegraphy to Television* (Durham, NC.: Duke University Press, 2000).
15 Hanns Eisler and Theodor Adorno, *Composing for the Films* (London: Athlone, 1994 [f.p.1947]), 75–6; K. J. Donnelly, "The Ghostly Effect Revisited," in *The Routledge Companion to Screen Music and Sound*, ed. Ron Sadoff, Miguel Mera and Ben Winters (New York: Routledge, 2017), 17–25.
16 Paul Newland, "Folk Horror and the Contemporary Cult of British Rural Landscape: The Case of *Blood on Satan's Claw*," in *British Rural Landscapes on Film*, ed. Paul Newland (Manchester: Manchester University Press, 2016), 162; Peter Hutchings, "Uncanny Landscapes in British Film and Television," *Visual Culture in Britain* 5, no.2 (2004): 29.
17 James Duncan, *The City as Text: The Politics of Landscape Interpretation in the Kandyan Kingdom* (Cambridge: Cambridge University Press, 1990), 17.
18 Alexander Porteous, *The Forest in Folklore and Mythology* (London: Macmillan, 1928); Bruno Bettelheim, *The Uses of Enchantment: The Meaning and Importance*

of Fairy Tales (London: Penguin, 1991); Elizabeth Parker, *The Forest and the EcoGothic: The Deep Dark Woods in the Popular Imagination* (New York: Palgrave, 2020).

19 See further discussion of this in Louis Bayman and K. J. Donnelly, eds., *Folk Horror: Return of Britain's Repressed* (Manchester: Manchester University Press, 2023).

20 Christian Norberg-Schulz, *Genius Loci: Towards a Phenomenology of Architecture* (New York: Rizzoli, 1980); Phil Legard, "The Haunted Fields of England: Diabolical Landscapes and the Genii Locorum," in *Folk Horror Revival: Field Studies*, ed. Katherine Beem and Andy Paciorek (Durham: Wyrd Harvest Press, 2015).

21 Korven taught himself many exotic instruments and has had a career involving both pop and jazz. He wrote the music for Patricia Rozema's acclaimed *I've Heard the Mermaids Singing* (1987) and *Cube* (1997) before *The VVitch*.

22 Perhaps less surprisingly, this instrument was also used in Bobby Krlic's score for Swedish-set folk horror film *Midsommar* (2019).

23 Mark Korven, "We Interviewed Composer Mark Korven about His Terrifying Score for 'The Witch,'" by Jonathan Barkan, *Bloody Disgusting*, February 26, 2016, https://bloody-disgusting.com/interviews/3381360/we-interviewed-composer-mark-korven-about-his-terrifying-score-for-the-witch/ (accessed December 20, 2021).

24 Jacob Stolworthy, "*The Witch* Composer Mark Korven: 'I Was Pleasantly Surprised the Film Did So Well,'" *The Independent*, Friday July 1, 2016, https://www.independent.co.uk/arts-entertainment/films/features/the-witch-composer-mark-korven-interview-robert-eggers-horror-black-phillip-philip-glass-a7113236.html (accessed November 10, 2017).

25 "*The Witch*: Behind the Music—Part 2," https://www.youtube.com/watch?v=Vdq1LA-PKRw (accessed December 20, 2021).

26 Ibid.

27 Some of Ligeti's music is also used in Kubrick's *The Shining*, too, which Eggers claims was a major influence on *The VVitch*. Jacob Hall, "The Influences of *The Witch*, Part One: Director Robert Eggers on *The Shining*," at */Film*, February 16, 2016, https://www.slashfilm.com/the-witch-influences-the-shining/2/ (accessed April 20, 2020).

28 Some are noted in Anton Bitel, "Diabolical Dualities in Robert Eggers' *The Witch*," in *The VVitch: A New England Folk Tale*, book with limited edition Blu Ray of *The Witch* (UK: Universal/Second Sight Films, 2022).

5

Producing Paranormal Sounds: Electronic Music, Projection, and Blurred Boundaries in *The Legend of Hell House* (1973) and *The Stone Tape* (1972)

Jamie Sexton

This chapter will interrogate the use of electronic sounds to denote ghostly and occultic phenomena in the BBC television film *The Stone Tape* (Sasdy, 1972) and the feature film *The Legend of Hell House* (Hough, 1973). Both films were made in the early 1970s, feature scientific investigations of the paranormal, and included electronic scores by musicians associated with the BBC Radiophonic Workshop:[1] Desmond Briscoe and Glynis Jones created *The Stone Tape* score for the Radiophonic Workshop, while Delia Derbyshire and Brian Hodgson, who had recently left the Workshop, produced the score for *The Legend of Hell House* for the newly established Electrophon Studios.[2] While science fiction was the genre most linked to electronic sounds around this period, such sounds also increasingly accompanied other types of filmmaking and programming, including horror, where the unusual sonic timbres of electronic music were often utilized for their contribution to a sense of alterity and fear. Nevertheless, extensive electronic scores were still rare for horror productions when these films were released. Further, I will argue that these films can both be connected to some "scientific" approaches to paranormal phenomena that were in circulation when they were made, such as T. C. Lethbridge's ideas about ghostly encounters, and the emerging practice of Electronic Voice Phenomena (EVP), which claimed to detect voices of the dead in recorded messages. Both these films engage with science and the paranormal and in doing so stress oppositions between them, but they also question such divisions (through, for example, highlighting the mysterious and occult nature of science). This logic of undermining categorical distinctions also extends to other themes and practices, such as the ways that both films blur temporal borders and undermine divisions between music and sound effects.

The BBC Radiophonic Workshop and Electronic Music

The BBC Radiophonic Workshop has become increasingly lauded and inserted into histories of electronic music. It was established in 1958 to create special sounds for radio and television. While in its early years it created mostly effects, it would over time produce full scores (as is the case with *The Stone Tape*). Louis Niebur has noted that, while in its early years the Workshop was mostly creating *musique concrète*-inspired sound via tape editing, over time it began to also produce "works that are driven primarily by tape-loop-created rhythmic patterns and finally to a combination of these rhythmic structures and tonally based composition."[3] While we cannot reduce these developments to purely technological causes, the Radiophonic Workshop's adoption of newer technologies during its existence was an important factor. Earlier Workshop output was primarily tape-based, though also made use of electronic sound generators. In the late 1960s, however, the Workshop started working with synthesizers which would henceforth become their most important compositional tools, though it did not abandon other approaches. When *The Stone Tape* was made, the Workshop had use of EMS VCS 3s and the more sophisticated albeit unwieldy EMS Synthi.[4] Hodgson and Derbyshire, meanwhile, also had EMS synthesizers within their Electrophon studios.

The Radiophonic Workshop created a large range of radio and television sounds throughout the 1960s and 1970s, and some of its members—whether current or former—would go onto create sounds for productions outside of the BBC. Examples include Desmond Briscoe's sound effects for the haunted house classic *The Haunting* (Wise, 1963). Briscoe also created effects for *Children of the Damned* (Rilla, 1964), *The Ipcress File* (Furie, 1965), and *The Man Who Fell to Earth* (Roeg, 1974), and is credited with "additional electronic music" for Saul Bass's sole feature as director, *Phase IV* (1974). Delia Derbyshire and Brian Hodgson, meanwhile, produced two albums for music production libraries under pseudonyms: with David Vorhaus they recorded *Electronic* for Standard Music Library in 1969, and with Don Harper they produced *Electrosonic* for the KPM library in 1972; the majority of tracks they recorded for Standard would eventually be used in Thames Television's science fiction series *The Tomorrow People* (1973–9).⁵ Daphne Oram, who was a co-founder of the Workshop despite leaving within a year, also contributed electronic effects for the ghost film *The Innocents* (Clayton, 1961).

Despite the presence of electronic sound within British television and radio, much electronically produced music and sound was still, broadly, perceived as strange and unpleasant by many critics.⁶ The strangeness of electronic sounds led to their incorporation into science-fiction films and programs more than other genres, where the alienness of electronic music was seen to complement the representation of alien creatures, advanced technologies, and distant worlds. Nevertheless, there were still few soundtracks that featured completely electronic scores outside of *Forbidden Planet*⁷; the most common tactic was to incorporate electronic sounds into existing scoring conventions, where they would often supplement orchestral music, adding exotic color to scores. The theremin, patented in 1928, was one of the first electronic instruments, and tended to be used in such ways, where it would often signify eeriness. There was nevertheless a gradual increase in the use of electronic instrumentation within screen media, influenced by a combination of people becoming more accustomed to new sonic textures and a growth in the availability of electronic instruments, particularly synthesizers. However, electronic sounds in the 1960s were still largely used as sound effects (as opposed to music): Briscoe's effects for *The Haunting* and Oram's for *The Innocents*, which were not credited, supplemented the music scores by Humphrey Searle and Georges Auric respectively, while the soundtrack for *The Forbidden Planet* was originally created as effects and credited as "electronic tonalities" rather than "music."⁸

Electronic Sounds in *The Stone Tape* and *The Legend of Hell House*

When *The Stone Tape* and *The Legend of Hell House* went into production, electronic sounds still tended to connote strangeness, eeriness, fear, and terror, which led to their employment within a few horror and thriller productions. As horror productions with complete electronic scores were still a rarity, the composers would have had to think carefully about what types of sounds were suitable for the productions they were scoring. Music/sound production, like many other processes, exists within a broader field of sonic conventions, and despite the novelty of these soundtracks, they nevertheless drew on previous electronic music scoring *and* non-electronic horror music/sound. In discussing the sounds of Gothic horror, Isabella van Elferen has argued:

> Gothic music exploits sound's ambiguous relation with embodiment, pushing the uncanny implications of this relation to their limits. Timbres like that of the "spectral" high-pitched violin, of the "transcendent" female choir, or of white noise suggesting "the ghost in the machine" are privileged within the genre, whether they are described in Gothic novels, heard in film and television, or interacted with in video games and on the dance floor. Musical elements undermining closure, such as the open-ended glissando and the repetitions of drones and non-linear music, increase the sense of uncanniness in sonic liminality.[9]

Both soundtracks continue some of the sonic traits that van Elferen identifies as common within the Gothic genre. They both, in different ways, incorporate drones and deploy nonlinear music, therefore conforming to some of the sonic tropes previously established not only within the Gothic genre, but also in other areas such as science-fiction, and occasional thriller material. While *The Stone Tape* does use drones, often in conjunction with electronically treated voices, *The Legend of Hell House*'s score draws on established ghostly sonic formulae to a greater extent. Its more conventional nature is evident in its employment of some acoustic instrumentation; the main theme, for example, features a pipe organ, which is a recurring sonic trope of horror films, especially Gothic horror.[10] This is supplemented by trumpet, bass clarinet, and an electronic drum track (both the latter are low-frequency sounds indicative of an ominous menace). Electronic sounds in the film, while texturally innovative, do employ high-pitched, airy drones and motifs, which have often been drawn upon to signify ghostly representation. For example, the partial employment of sounds which denote ghostly presence resembles the sounds of wind or breath. These sounds—often quiet in volume—form a sonic frame denoting the eerie atmosphere of the haunted house, fuelled by a malevolent entity. There is also a slightly sinister, two-tone electronic motif that recurs throughout the film as a kind of punctuation marker. This occurs when there is a temporal ellipsis and is accompanied by the date and time being announced via titles. These dark and spectral sonic markers are occasionally displaced by louder, more dramatic and dissonant sounds that erupt when Belasco's spirit attempts to attack the guests.

Contrastingly, *The Stone Tape*'s soundtrack, while at times drawing on some familiar sonic tropes, is stranger and functions more sparsely overall. Briscoe and Jones's score foregrounds synthetic sounds, though it does also employ drones and manipulated voices. Referring back to Niebur's point about the development of Radiophonic sound, *The Stone Tape* combines *concrète*-style effects with loop-based motifs. Its main theme is a descending ostinato of detuned, sustained tones, which acts as a recurrent motif through the film, and which also forms the basis of the main theme used over the credits. This looped motif creates an eerie, unsettled atmosphere and fittingly connotes the strange goings-on within an isolated room of the research building. It is, nevertheless, slightly odd for such music to accompany a horror film, though its scientific and technological themes can be considered an important factor in their deployment. The descending motif also connotes other themes of the film. The looped nature of it relates to how time itself is looped within the film, while the downward trajectory of the pattern refers to the idea of descending back in time, even to a hellish region (further emphasized by prominent red lighting when characters walk down the corridor toward the haunted room).

Ambiguity, Sound, and Projection

Both films involve the scientific investigation of paranormal phenomena. While science and the paranormal are often differentiated, there have been some notable overlaps between the disciplines.[11] One of the most significant was Spiritualism, which became popular in the Victorian era and influenced several novels and films. Murray Leeder has explored the influence of Spiritualism on *The Legend of Hell House*, arguing that the supernatural and the scientific are blurred in the film (and novel), which is indicative of a society "simultaneously committed to rationalism and obsessed with the supernatural and the occult."[12] While I agree with Leeder about the influence of Spiritualism on the film, I think it is also crucial to keep in mind some of the perspectives on supernatural phenomena that were circulating nearer the time these respective films were produced. One figure whose theories can be considered important—particularly within *The Stone Tape* but to an extent in *The Legend of Hell House*—is T. B. Lethbridge. Another tradition I want to draw on is the phenomenon of Electric Voice Phenomena (EVP). The latter is particularly important as it is not addressed beyond a brief mention in Leeder's article, which largely focuses on the "optical uncanny."[13]

The concept of projection has been used to theorize ghostly phenomena at least since the late nineteenth century,[14] but in the theories of parapsychologist T. C. Lethbridge it was linked to the idea of *recordings*. Firstly, he proposed the idea that ghosts themselves were recordings, and he has sometimes been credited as originating the "stone tape theory."[15] He argued that any traumatic event could project energy onto rocks and other matter, which store these events and, in certain circumstances, can replay them. For Lethbridge, ghosts were recorded matter and could be detected by observers who had powers of extra-sensory perception (ESP). He conceived of human ESP through metaphors of machinery, including recording. He argued that the brain makes recordings and that these "are made available by a limited number of senses, each with a limited range," but that "persons whose extra-sensory perception is high, appear to be able to register other ranges on their machines."[16] Lethbridge considered the human brain as a machine that the separate—and seemingly non-machinic—mind was often unable to affect. People with higher levels of ESP were capable, in his terms, of undoing a screw in their minds and "altering their mental 'voltage.'"[17]

EVP refers to voices heard on recordable media that are not audible to perceivers at the time of the recording, leading some figures to claim that such voices are evidence of dead spirits attempting to communicate with the living. While he was not the first person to propound such ideas, Konsantin Raudive popularized theories of EVP with his 1968 book *Unhörbares Wird Hörbar* (The Inaudible Becomes Audible), which was translated into English in 1971 under the title *Breakthrough: An Amazing Experiment in Electronic Communication with the Dead*. The British publication coincided with a separate release on vinyl, which featured several EVP recordings that Raudive had collected.[18] The EVP recorded by Raudive, like subsequent EVP audio, is characterized by, in Joe Banks's words, "brief bursts of extremely distorted sounds and voices, accompanied by what appears to be motor noise picked up from the recording apparatus by a microphone."[19]

Banks does not believe that EVP offers evidence of ghost voices. He argues that the conflux of sounds within such recordings is *perceived* by many to be dead voices because they are cued to interpret them as such. That is, many people *project* their own desires and beliefs onto the sonic data. As these recordings were low-fidelity, they were subject to greater extraneous noise which, for Banks, resulted in a greater likelihood that spirit voices would be detected through an "aura of menacing low-fidelity mystique, which can help impart a subjective impression of authenticity to such material."[20] Voices caught on high-fidelity recordings, by contrast, do not produce the same convincing results. Banks largely debunks EVP recordings as evidence of actual spirits, though takes them seriously as a psychological phenomenon. He argues that creating EVP recordings is "child's play," merely requiring primitive recording and overdubbing techniques. We can link the low-fidelity EVP recordings to broader ideas around recording and space, through a consideration of Alvin Lucier's legendary 1969 sound artwork "I Am Sitting in a Room." Lucier recorded a short passage of him speaking on tape, then re-recorded onto another tape, a process repeated multiple times. Lucier's work creatively illustrates how the materiality of recording media, and the specific spatial location in which recordings are made, impact upon the production. The constant repetition of the recording process eventually amplifies the frequent resonances of the recording space while rendering the spoken text unintelligible.

Lucier's experiment has already been drawn on by Daniel Pieterson to analyze *The Stone Tape*. Pieterson uses the term "spectral resonance" to refer to how layers of "hidden" audio information accrete through each re-recording, which can transform the recorded piece to such an extent that the original recording itself becomes a ghostly presence.[21] Pieterson notes how *The Stone Tape* features an attempt by the researchers to unlock the stone recording so that it can be played on demand, but which unwittingly "wipes" the recording. Sound, space, and time become very important within *The Stone Tape*, just as they were in Lucier's work (albeit in very different ways). While the passage of time will degrade an original recording, such degradation can also allow new sonic elements to form, although these might not be comprehensible to human perceivers. In *The Stone* Tape, while the recording is unwittingly "wiped" there are nevertheless other sonic traces encountered by computer specialist Jill (Jane Asher), which stretch back farther in history, and which are particularly dense in texture, featuring multiple layers of reverberant noise.

Projection and "spectral resonance" should be linked to the more abstract sounds employed within both *The Stone Tape* and *The Legend of Hell House*. Audio characterized by dense textures and/or low-fidelity recording techniques will more likely prove difficult to understand for many listeners, whether presented as music or not. Abstract music has, as is well known, often faced hostility from more traditional musicians and listeners, who consider it noise, while low-fidelity recordings tend to muddy audio information, resulting in greater ambiguity. Both will tend to require more mental work from a perceiver who wants to understand/interpret them: abstract music because it differs from more conventional musical modes and may subsequently require greater effort to comprehend, low-fidelity recordings—and particularly those which involve re-recording—because they introduce noise via low signal-to-noise ratio. What is considered noise by one person may, however, be perceived as something entirely different by other listeners.

In their cataloguing of "sonic effects," Jean-François Augoyard and Henry Torgue emphasize how there "is no universal approach to listening" as sound cannot be "isolated from the spatial and temporal conditions of its signal propagation" and because "sound is also shaped subjectively; depending on the auditory capacity, the attitude, and the psychology and culture of the listener."[22] They also stress how perception of sounds becomes more difficult in low-fidelity sound environments, due to their "blurred and hazy" nature.[23] Such spatial and subjective dimensions of sound, in addition to the difficulties of interpreting certain sounds, are apparent in *The Stone Tape* and *The Legend of Hell House*. That both films employ electronic sounds and combine score and effects, making frequent use of abstract drones, is important in heightening the ambiguity of sound and foregrounding the subjective effort involved in the interpretation of such sounds. Fittingly, different characters will interpret these (largely) sound events in different ways. While in *The Legend of Hell House* this is connected to the different intellectual attitudes of the respective characters, in *The Stone Tape* there is greater emphasis on the subjective ways that people respond to the sounds and other ghostly phenomena. Some people do not hear the sounds that others hear, for example, while character Stewart (Philip Trewinnard) cannot feel the cold that others experience in the room.

Differences in interpretation can further be metaphorically linked to abstract and/or electronic music within this context: the novelty of some of the timbres, combined with deviation from previous music, made it more difficult to interpret such sounds due to the absence of perceptual templates geared toward understanding them. Shots in both films where characters experience sonic assaults—where the sounds become heightened in volume and frequency—lead to many shielding their ears in agony, a reaction that many opposed to experimental electronic music would have sympathized with (such shots are more frequent within *The Stone Tape*). It should be noted, however, that while I am focusing on sound, it is still important to consider visual aspects of the respective films, as other sensory information influences how we hear.

Blurred Boundaries

As noted, both films employ electronic scores that combine sound effects and music, which are conventionally created by different teams. K. J. Donnelly has highlighted the growing integration between the categories of sound effects and music, influenced by the digitization of both realms, so that effects are now more likely to be treated musically, while music increasingly is conceived in sonic terms. Donnelly argues that these integrated soundtracks lead to a "notable collapse of the *space* between diegetic sound and non-diegetic music."[24] The ways such composite scores can question fixed boundaries (in this case diegetic and non-diegetic) further relate to how both films more broadly collapse, or at least trouble, categorical demarcations.[25]

The first type of opposition to be interrogated is between science and the paranormal. As this theme has been discussed, I will not pay too much further attention to it. Both films of course draw on existing traditions of scientific research into the paranormal, as well as various fictions that have involved such investigations. Yet they do not merely represent the blurring of oppositional boundaries; they also present science (and technology) as mysterious, with its own occulted blind spots (that is, phenomena that cannot be adequately explained rationally). Computational machines are key tools in the scientific investigations of ghostly presences, and in both films are sometimes presented as abstract and mysterious. In *The Legend of Hell House*, for example, the mise-en-scène—and a significant portion of the soundtrack—signals the Gothic, but the arrival of Dr. Barrett's (Clive Revill) scientific machine disrupts this Gothic facade. Its stark contrast with its surroundings renders it a strange presence, an almost mystical machine.[26] *The Stone Tape* is different in that it embeds Gothic elements—most notably the exterior of the house and the haunted room that is left unrenovated—within a more clinical, modern setting.[27] Yet *The Stone Tape* also alludes to the occult underbelly of science. This is most prominently indicated at the beginning, including the brief credits sequence. The credits display a sequence of computer imaging—moving green abstractions—accompanied by slightly dissonant, electronic tones underpinned by a metronomic ticking sound. After the title is displayed, there is a cut to an incredibly blurred image and a disarming, manipulated guttural sonic churn in conjunction with the metronomic ticking that has continued from the credits, intercut with an extreme close-up of Jill's eyes. The camera racks focus slowly to eventually reveal futuristic lettering of "Ryan Electronics" on the back of a removal truck. This scene also establishes Jill as someone who has special mental powers, who can sense things that lie beneath normal exterior perception, as she senses this before prior to arriving within the grounds of the research center (when she does park within the grounds of the center, she is then alarmed by two Ryan Electronics removal trucks which reverse toward her, causing a momentary sense of panic which indicates that she has premonitory powers).

While it is not clear at this early stage in *The Stone Tape*, the demonic-like sounds and blurred visuals are ultimately symbolic of the slippages between past and present. Such slippages are, of course, central in media narratives about ghosts, but *The Stone Tape* does emphasize the ways that the past can manifest itself in the present beyond a singular apparition from a deceased individual, and the soundtrack tends to reinforce such ideas. Such overlaps are most notable in a visual sense: the research center, for example, is mostly marked by modern features, particularly the banks of computers, even though it is situated in an older building. However, one room is left unrenovated because the workmen believe it to be haunted. This room is visually contrasted to other parts of the building. It contains an old wall and is shot in very low light, which lends it a tenebrous feel, in contrast to the white, brightly lit nature of all other rooms. We also discover—via Collinson—that the room is older than other parts of the building, which was built around it, and that it could possibly stem back to the Saxon era. The stark contrast between these spaces emphasizes differences between the past and present in one sense, but the room's existence within the same building as the other rooms signals how they also bleed into one another: the room is a literal example of how the past always leaves traces which are effective in the present. The sound of the film loosely indicates past and present via different sound production techniques. While *musique concrète* was still relatively new, within the field of electronic composition it was nevertheless increasingly rendered an antiquated approach in the age of the synthesizer. *The Stone Tape* tends to associate the sounds of the haunted room, at least initially, with older, more earthy-sounding textures, including manipulated vocal sounds. In contrast, the world of scientific research is often associated with synthesized descending loops.

The Legend of Hell House does not merge these categories to the same extent, though as a haunted house film it inevitably contains some slippages between past and present. The exterior shots, for example, were filmed at Wykehurst Place. As a Gothic revival building, its actual status already denotes temporal mixing. If the design of the interior is also more traditionally Gothic in appearance, it is occasionally interrupted by more modern paraphernalia such as Barrett's scientific machine. Score and effects, at times, seem to be more easily to distinguish than in *The Stone Tape*, with electronic effects being used to represent the absent presence of Belasco via airy sounds, and the dramatic impact of objects clanging and bashing against material surfaces during Belasco's attempts to terrorize the inhabitants. Contrastingly, the credits feature more conventionally musical sounds and employ some nonelectronic forms of instrumentation.

Conclusion

Both *The Stone Tape* and *The Legend of Hell House* should be considered important horror soundtracks for their sustained employment of electronic sounds and merging of score and sound effects. Niebur states that the advent of synthesizers led the Radiophonic Workshop gradually away "from electronic music's distinctly avant-garde roots," partly because of the "rising acceptance of electronic sounds, gradually becoming comfortably familiar to audiences."[28] He does qualify this argument, though, by stating how this shift did not occur overnight, and that "even with the arrival of synthesizers, there was still a marked tendency on the part of several composers to associate electronically generated sound with atonality."[29] These films were made at a time when synthesizers were still quite new, when their parameters and possibilities had not yet been fully explored. When the BBC obtained an EMS Synthi, it did not even come with a manual, hence there would have been an experimental phase of discovery and exploration of these machines during the late 1960s and early 1970s.[30]

The experimental and atonal textures of these respective soundtracks, however, are also influenced by how they functioned within narratives based around hauntings and fear. Many of the sounds used are eerie and, at moments of high tension, dissonant. In this sense they marshal sound's power to generate fear and anxiety, as outlined by Steve Goodman who writes that sound "is often understood as generally having a privileged role in the production and modulation of fear, activating instinctive responses, triggering an evolutionary functional nervousness."[31] That these powerful, frightening, and unnerving sounds are mostly electronically produced also reflects these films' status as scientific horror, with electronic sounds still often at this stage heavily connected to science fiction productions.

Not only were electronic sounds often perceived as odd and perplexing at the time these films were made, the machines and people who created them also tended to be treated as slightly odd and mysterious. In a 1970 article for the *Guardian*, Kirsten Cubitt wrote an article about the Radiophonic Workshop which opened with the following comments:

> It struck me immediately as the contemporary equivalent of an alchemist's kitchen; to my lay eye the BBC radiophonic workshop in Maida Vale took shape as an esoteric, slightly scary clutter of instruments, some quite dusty, others unintelligibly new, with fussy, jiggling dials, and a running commentary of gulping noises, and bumps, and whinnies.[32]

This mysterious element of electronic music, and advanced technologies more generally, informs the ambiguity of sonic messages within the films (while there are some visual factors at play in detecting ghostly presences, sound acts as the primary means of communication between the dead and living). Both films feature different people trying to decode sonic information: in *The Legend of Hell House* the interpretive frameworks of three people—a scientist, a mental medium, and a physical medium—compete to explain the supernatural occurrences at the house. In *The Stone Tape* those attempting to understand the goings on in the haunted room are scientists, but they still differ over their explanations, with gender a factor fuelling arguments. Peter, the director of the research team, is a control freak who thinks he can explain and master all before him. His arrogance leads to him wiping the tape and, ultimately, to Jill's death. While Jill is shown as extremely intelligent and analytical, she is often dismissed by Peter as being over-emotional, despite her arguments being more intellectually rigorous than his.[33] This sexist attribution again links to Peter's overconfidence in being able to understand the world around him, for he wrongly interprets her sensitive nature. He assumes it is indicative of a highly strung female, and therefore is a hindrance to their investigations, but her sensitivity is more related to being attuned keenly to her surrounding environment than it is to emotional sensitivity. She would, for Lethbridge, be considered a person with extra-sensory perception.

In both productions, experimental electronic sounds connote a wide range of perceptual reactions to strange sounds, and these are often associated with different fears: fear of ghosts and demonic activities, fear of the female's rising status in society, but most of all fear of the unknown, and of the unknowable, which can often draw attention to the limits of human knowledge and point to the blind spots of scientific research. Discussing horror and philosophy, Eugene Thacker has written that horror often confronts the very limits of human beings' ability to "adequately understand the world at all," a process that is central to the multifaceted set of fears that fuel both *The Stone Tape* and *The Legend of Hell House*;[34] even though their narratives work toward resolution, the void of unintelligibility—as well as the limits of human frameworks to arrive at certainty—is perhaps the most frightening specter haunting the narrative of both productions.

Notes

1 The scores are composite in the sense of combining the conventionally separate categories of music and effects (as I will discuss later in the chapter). They are almost completely electronic, but there are some exceptions to this. In *The Stone Tape*, there is one more conventional piece of contemporary pop music—"Scene and Heard," composed by Paddy Kingsland (also a member of the Workshop)—that plays diegetically from a car. In *The Legend of Hell House* there are a few bits of acoustic colour in parts, most notably the opening theme.
2 Hodgson set up a studio, Electrophon with composer John Lewis and Derbyshire, though Derbyshire would leave soon after. See Louis Niebur, *Special Sound: The Creation and Legacy of the BBC Radiophonic Workshop* (Oxford & New York: Oxford University Press, 2010), 43.
3 Ibid., 65.
4 The first commercial synthesizer—the Moog—became available in 1964 while more portable and affordable synthesizers followed, such as the EMS VCS 3 (1969) and the Minimoog (1970). This led to the increased use of electronics within pop and rock music in the 1970s, as well as a slight increase in electronic music being used in films. According to Briscoe and Curtis-Bramwell, the BBC Radiophonic Workshop were working with the VCS 3 officially in 1968. As this was before they became generally available, it may be that they were using a prototype model. Niebur has noted that Derbyshire and Hodgson were using a VCS 3 prototype for their work as Delta Unit One from 1967. Niebur claims that the Workshop did not actually purchase a VCS 3 until 1970. Soon after they purchased the more complex (and expensive) EMS Synthi 100 (also known as the "Delaware"). Frances Morgan has claimed that the BBC Radiophonic Workshop "was the first recipient of a

finished Synthi 100." See, respectively, Niebur, *Special Sound*, 131; Desmond Briscoe and Roy Curtis-Bramwell, *The First 25 Years of The BBC Radiophonic Workshop* (London: BBC, 1983), 128; Frances Morgan, *Electronic Music Studios London Ltd (EMS), the Synthi 100 Synthesizer and the Construction of Electronic Music Histories* (PhD thesis, Royal College of Art, 2021), 154.

5   Derbyshire was credited as Li De La Russe and Hodgson as Nikki St. George on both records. Many musicians would adopt pseudonyms when they produced library music to evade contractual issues.

6   Joanna Demers, *Listening Through the Noise: The Aesthetics of Experimental Electronic Music* (New York: Oxford University Press, 2010), 21.

7   Hitchcock's *The Birds* (1963) is one other example, though the films did use these electronic sounds quite sparsely and they would tend to be largely categorized as sound effects (in contrast to the source music featured in the film). The electronic sound was created by Oskar Sala and Remi Gassmann, primarily on a trautonium that Sala had developed. Another example of a fully electronic score was Pierre Henry's music for *Maléfices* (Decoin, 1962).

8   Bebe and Louis Barron were not hired to score the film. David Rose did produce a musical score, but this was not used in the finished film See Reba Wissner, "A Universal Mind: The Film Music of Bebe Barron," in *Women's Music for the Screen: Diverse Narratives in Sound*, ed. Felicity Wilcox (New York: Routledge, 2022), 28.

9   Isabella van Elferen, *Gothic Music: The Sounds of the Uncanny* (Cardiff: University of Wales Press, 2012), 4.

10 Julia. Brown, "*Carnival of Souls* and the Organs of Horror," in *Music in the Horror Film: Listening to Fear*, ed. Neil Lerner (New York: Routledge, 2010), 5.

11 See, for example, Jeffrey Sconce, *Haunted Media: Electronic Presence from Telegraphy to Television* (Durham: Duke University Press, 2000) and Richard Noakes, "Spiritualism, Science and the Supernatural in Mid-Victorian Britain," in *The Victorian Supernatural*, ed. Nicola Brown, Carolyn Burdett and Pamela Thurschwell (Cambridge: Cambridge University Press, 2004), 23–43.

12 Murray Leeder, "Victorian Science and Spiritualism in *The Legend of Hell House*," *Horror Studies* 5, no.1 (2014): 33.

13 This phrase is from Tom Gunning. See Gunning, "Uncanny Reflections, Modern Illusion: Sighting the Modern Optical Uncanny," in *Uncanny Modernity: Cultural Theories, Modern Anxieties*, ed. Jo Collis and John Jervi (Houndsmills: Palgrave, 2008).

14 See Murray Leeder, "'Visualizing the Phantoms of the Imagination': Projecting the Haunted Minds of Modernity," in *Cinematic Ghosts: Haunting and Spectrality from Silent Cinema to the Digital Era*, ed. Murray Leeder (New York: Bloomsbury, 2015).

15 Lethbridge did not originate such theories, which can be traced back to "place theory" and psychometry. His book sold well, though, so he is often credited with the popularization of such ideas. Nigel Kneale has claimed he came up with the

Stone Tape idea himself, though he may have been unaware of such ideas as they had been in circulation prior to this. See Nigel Kneale and Kim Newman, Audio Commentary for *The Stone Tape*. *The Stone Tape* (BFI DVD, 2001).

16 T. C. Lethbridge *Ghost and Ghoul* (London: Routledge and Kegan Paul, 1967), 136.

17 Ibid.

18 The recording has been reissued a few times since the 1971 Vista release, which was issued on 7" vinyl (with a playback speed of 33 1/3). In 1982 *The Unexplained* magazine issued the same content on a flexidisc that came with issue 1, while the complete contents of the disc were also included, alongside several other examples of EVP, on the CD *The Ghost Orchid* (1999).

19 Joe Banks, *Rorschach Audio: Art and Illusion for Sound* (London: Strange Attractor, 2012), 14.

20 Ibid., 15–16.

21 Daniel Pieterson, "I Am Sitting in A Room: Spectral Resonance as a Source of Horror in the Work of Alvin Lucier and Nigel Kneale's The Stone Tape," in *Folk Horror Revival: Harvest Hymns 1, Twisted Roots*, ed. Jim Peters, Richard Hing, Grey Malkin and Andy Paciorek (London: Wyrd Harvest Press, 2018), 273.

22 Jean-François Augoyard and Henry Torgue, *Sonic Experience: A Guide to Everyday Sounds*, trans. Andra McCartney and David Paquette (Quebec: McGill-Queen's University Press, 2005), 4.

23 Ibid., 7.

24 K. J. Donnelly, *Occult Aesthetics: Synchronization in Sound Film* (New York: Oxford University Press, 2014), 126. Donnelly argues that this is becoming more common, but does note earlier examples, such as *Forbidden Planet* and *The Birds*.

25 The Radiophonic Workshop itself straddled boundaries in that it was separate from the music department but developed into a unit that was not confined to just sound effects.

26 That the apparatus is only referred to as a scientific machine heightens the sense of its mystery.

27 Strictly speaking, the exterior building used for location filming, Horsley Towers, was Neo-Tudor, though it had additional structures added to it in neo-Romanesque, which is often considered a subset of Gothic architecture. Ada Lovelace, the pioneering scientist linked to the development of the first computer program, used to reside at Horsely Towers, which links also to the general theme of scientific research, the importance of computers, and of a brilliant female scientist.

28 Niebur, *Special Sound*, 125.

29 Ibid.

30 Frances Morgan notes that Brian Hodgson himself wrote a manual for the Synthi 100. See Morgan, *Electronic Music Studios London Ltd (EMS), the Synthi 100 Synthesizer and the Construction of Electronic Music Histories*, 163.

31 Steve Goodman, *Sonic Warfare: Sound, Affect, and the Ecology of Fear* (Cambridge, Massachusetts: MIT Press, 2010), 65.
32 Kirsten Cubitt, "Dial a Tune," *The Guardian* (September 3, 1970): 9.
33 Peter acknowledges Jill's intelligence, which is why he wants her on the team, but his appreciation of her intellect is often countered by his sexism; for example, his plea to her that "I need you for your brain" follows him stating "oh my Jilly, you're a very female one" when he finds her crying.
34 Eugene Thacker, *In the Dust of This Planet: Horror of Philosophy Vol. 1* (Alresford: Zero Publishing, 2011), 1.

6

Cries and Whispers: Landscape and Sound in *The Owl Service* (1969) and *Red Shift* (1978)

Craig Wallace

This chapter explores the relationship between landscape and sound in television adaptations of Alan Garner's novels *The Owl Service* (Peter Plummer, 1969) and *Red Shift* (John Mackenzie, 1978), the former an eight-part series made for Granada Television and the latter a single play for the BBC *Play for Today* strand. In *The Owl Service*, the sounds of scratching on the ceiling, the flapping of wings, and a motorcycle engine associated with legends and events in the landscape of the "past" are heard in the "present." In *Red Shift*, protagonists from the Roman occupation of Britain, the English Civil War, and from the 1970s experience the landscape simultaneously. There is no sense of chronological distance between them, and this is conveyed through the interaction of editing and sound. The landscapes are, in a way, haunted by ghosts, but not simply of the past, present, or future. Instead, the sonic hauntings are multidimensional, perceptible at once. Sounds are linked with place, and shared by those that inhabit the landscape, rendering the conscious experience of the environment in seemingly different times as contemporaneous. The series and play use textual elements of the television ghost story genre, such as archaeological excavation and eerie sounds heard in the settings. However, there is not a sense of a disruptive, repressed, and buried revenant returning from the "past" to haunt the "present." Sounds appear in landscapes conceptualized as archaeologically stratified and aligned with interior levels of consciousness.

In an archaeological conceptualization of time, scattered artifacts contained within a cross-section of the compacted strata of the earth are perceived concurrently, disrupting settled accumulated layers. Penelope Lively, a novelist contemporary with Garner, writes of the presence of the past in a landscape, where temporal moments cluster to form a constantly changing pattern through generational occupation and habitation.[1] This jumbling of time also applies to the geological strata beneath the surface containing scattered archaeological evidence of settlement. Layers of stone are not arranged tidily in linear sequence: they fold and pleat, and "old" and "young" sit side by side.[2] This stratified rock is quarried to build on the landscape across generations, creating temporal complexity. Like the surface of the landscape and the archaeological strata beneath, ghostly noises weave, knit, fold, and pleat, and in the television adaptations of the novels, they manifest through sound waves and vibrations, cries, and whispers.

In "Inner Time," Garner describes the way that painful memory traces are successively enclosed creating a cumulative pattern of recorded files that, once activated, all "seem to be present simultaneously,"[3] and therefore "two intensely remembered experiences [...] will be emotionally contemporaneous, even though we know that the calendar separates them by years."[4] This disrupts any sense of linear time, collapsing distinctions between the past and present. An analogy is made with the constellations in the night sky: "When we look at a starry sky, we see a group of configurations that seem to be equidistant from us and existing now. That is an 'apparent perspective'. We are looking at a complexity of times past, [...] all at different epochs, distances and intensities."[5] The co-existence of associated memory traces and connected emotionally painful experiences combine to produce a patterned "constellation of pain,"[6] which includes, Garner argues, the experiences of parents or grandparents genetically built in.

Garner describes how the production of *The Owl Service* adaptation on location ignored "concepts of time and space," filming out of narrative sequence with "no sense of dramatic progression [or] emotional development."[7] The novel is written in the past tense and the third person: out of sequence filming renders this distance gone and the time of the film is "now."[8] This is compared to the jumbled patterns and apparent perspectives of recalled memory, the disorientation of past moments made present simultaneously and the difficulty of reconciling the memory trace with the here and now: there is no "now" when the memory of a painful experience is in the present tense.[9] The realization of the novel in the filmmaking process is disorientating, with disparate moments or scenes shot out of sequence in the here and now. Linear time and narrative progression collapse in the exterior landscape that forms the setting of the novel, with sequences and scenes jumbled, forming an abstract pattern.

As an archaeologist Garner has excavated stone artifacts in the fields around his house close to the telescope at Jodrell Bank observatory. Garner unearthed flint tools from this landscape and his hand may be the first to hold them since the last hand 10,000 years ago. Generationally, the two hands are possibly linked.[10] Garner connects the intelligence that created the axe from flint and the dish of the radio telescope receiving signals or waves emitted from quasars or star-like objects.[11] The antennae is hearing "whispers" traveling at the speed of light and arriving at the telescope "Now": "if Now exists; if Now is possible."[12] When the hand "picked the pebble [to form the axe] the whispers had covered more than ninety-seven percent of their journey" to the site.[13] There is no such thing as now. Once excavated, the knapping sound produced by the flint tools is heard again in the landscape like an echo, collapsing linear time. When the sound is heard, linear time is nothing.[14] Receiving waves emitted from quasars, and viewing the stars in the night sky from the perspective of a translucent, stratified landscape, can be applied to the sound in the television adaptations of Garner's novels. The noises heard by characters in *The Owl Service* and *Red Shift* work on the same principle as the traveling waves of sound, the whispers emitted from quasars, creating an eerie effect.

The Owl Service

In *The Owl Service*, Alison (Gillian Hills), Roger (Francis Wallis), and Gwyn (Michael Holden) are on holiday in a valley in Wales where they discover an old dinner service decorated with a floral design. Alison traces the pattern and finds that the flowers can be rearranged to form paper owls. The valley setting, with hill, river, and standing stone, is connected to an ancient legend that appears in *The Mabinogion*, a fourteenth-century collection of stories that stem from an older oral tradition.[15] The story that forms the basis of *The Owl Service* is of Gwydion who conjures a wife for Lleu Llaw Gyffes out of flowers. Her name is Blodeuedd. She falls in love with Gronw Bebyr and together they plot to kill her husband. Lleu is killed by the bank of a river beneath a hill with a spear. After being restored by Gwydion he seeks redress, and Blodeuedd is transformed into an owl. Gronw faces a return blow from Lleu by the riverbank with a stone set between him and the spear, which pierces the stone and Gronw. Inhabiting "a mythic landscape,"[16] such as the settings of *The Mabinogion*, is to be "not trapped in linear time," but "in mythic time, where everything is simultaneously present."[17] The landscape jumbles temporal moments together and expresses the imaginative content of myth. Alison, Roger, and Gwyn, like Nancy, Bertram, and Huw of a previous generation, are drawn in to a recurring expression of the tragedy. Nancy's lover Bertram was killed in a motorcycle accident when Huw tampered with the brakes. Alison reads from *The Mabinogion* in the series and recalls the story to Gwyn and Roger, who make the connection between the story and the surrounding landscape, the hill, the river, and the stone with a hole in it: "we're right in the very place where all this flower power is supposed to have happened." The myth is expressed in the location through "the three who suffer every time."

The eight-part series, adapted by Garner and directed by Peter Plummer, was shot on film in color for Granada Television, and originally broadcast from December 1969 to February 1970. The exteriors were filmed on location in Dinas Mawddwy in Wales, where the novel is set. The patterning of elements of style such as framing, editing, and sound makes connections between the narrative and the legend, the characters and folklore types, and the characters and the landscape. The energy and power released in the valley location are represented through a combination of mise-en-scène, editing, and sound. For example, Alison, Roger, and Gwyn dash around a billiard table violently clashing the colored balls together. The colors correspond with their respective costumes: Alison in red, Gwyn in black, and Roger in green. Stephen McKay suggests that these colors are associated with each of the characters throughout the series, and that they also correspond with the "international wiring color code"[18] of the period. Later, in the same billiard room setting, the clashing of the three colors on the table, associated with electrical power, cause the matchboard wall paneling to peel and split, crumbling away to reveal a painting of a figure behind the surface. The painting depicts Blodeuedd framed by flowers and claws. Like the dinner service in the attic, the energy generated by the meeting of the three in the landscape causes artifacts from the apparent "past" to break through into the "present," excavating accumulated layers to be present simultaneously. The layers are perceivable at once, an abstract pattern of multiple iterations of the myth situated in the landscape: the dinner service, the wall painting, or archaeological artifacts excavated on the mountain, a spearhead and a slate pendant with an owl face pattern (both older than the written, recorded legend). The layering of the "past" over the "present" to convey a simultaneity is evident in the sound design in particular sequences in the series.

Ghosts appear in *The Owl Service* through an engagement with the sound and image of photography. Figures from myth slip and weave between layers of the stratified landscape, appearing on photographic prints taken with lengthy exposures on a delayed setting. Roger takes a series of photographs of the trees on the hill framed through the hole in the standing stone. When the film is developed, they show a figure by the trees, and when the prints are enlarged the figure becomes more distinct: a man holding a spear and then, later in the sequence, standing with a motorcycle. The photographs were taken within the space of two minutes using different exposure lengths. The apparitions captured by the camera on a delayed setting could be equated with traveling waves of sound, the whispers emitted from quasars. In a later episode Roger is posing for a self-portrait with the camera on delay. The "whirr" and "click" of the timer are heard on the soundtrack. The whirring sound is laid over a sequence of cuts between Roger and a close-up of the camera lens, back and forth as he retains the fixed pose for the exposure, until the click of the shutter is heard. The "whirr" and "click" of the camera mechanism, and the long exposure time and delayed setting, in which ghosts appear, work on the same principle as the waves of sounds from the "past" heard in the "present." Landscape, myth, and sound rupture a linear understanding of time.

The style of the opening title sequence makes a patterned use of shot composition, dissolves, and sound that function as narrative exposition, establishing motifs, and the unsettling tone of the series. The sequence consists of a series of still images of the landscape and animated sequences featuring shadow puppets, with hands forming the shape of birds and the hole of the Stone of Gronw. Variations of the musical theme, based on traditional Welsh folk songs performed on a harp, are repeated on multiple occasions throughout the series, and the traditional song "Ton Alarch" plays over the closing titles. In the title sequence the music is integrated with a number of sound effects heard diegetically in the series, such as a motorcycle throttle revving, the rapid fluttering of wings, and scratching noises, as well as the strange repeated sound of a bathtub emptying.[19] The sounds are disruptive and disturbing, and alongside dissolving images of the landscape, they express aspects of the legend in multiple iterations.

The sequence opens with the introduction of the harp theme played over an abstract design formed by dark slanted vertical stripes on a light background. The shot pulls out to reveal the trees on the hill, followed by several dissolves into images of the same landscape from the point of view of the bottom of the hill looking up. The camera position and angle have narrative significance, and the series of dissolves suggest multiple superimposed layers of time and space. A further dissolve transitions to a shot of the trees on the hill framed by a circular iris which begins to move away, revealing a series of diminishing concentric rings, suggesting circularity, reverberation, and depth. This shot is accompanied by the disconcerting sound effect sourced from a recording of a bathtub emptying, a clanging reverberation of water spiraling out through old pipes, suggestive of the rustic setting. This sound is repeated across the next three dissolves in the title sequence. The shot of diminishing circles is also accompanied by the intrusive and aggressive sound of a motorcycle throttle revving the engine. This sound is repeated at several points in the series, like a looped recording imprinted on the landscape, as multiple expressions of the myth are present simultaneously. This corresponds with the image of circularity and resounding depth.

The other sound effects used in the title sequence suggest further aspects of the myth. The animated shadow puppets use two hands to form the shape and motion of a flying bird, black on a bright white background as if the hands were backlit by a circular spotlight behind a translucent gauze screen. The transformation of human hands into the shape of a bird echoes the legend of Blodeuedd, and the loud and aggressive rustling of flapping wings is heard on the soundtrack. The puppet dissolves into a different shape, the hands forming a circle with the fingers and thumbs, like the hole in the Stone of Gronw that frames the trees on the hill, connecting the human inhabitants with the landscape and the legend. The next animated sequence depicts the pattern on the plates, gradually formed from an abstract owl head with eyes, brow, and beak, to a more complex floral design with stems and leafy heads. Three scratch lines suddenly appear, cutting vertically through the pattern leaving blank white marks, before wiping the image from the screen. The appearance of the scratch marks is overlaid with a scratching sound that is repeated across the transitions to the title cards. The sound, interspersed with brief moments of silence, is again disruptive and unsettling. The sound of the motorcycle, the flapping of wings and the scratching noises, integrated with the gentle sound of the harp, are audio motifs throughout the series, expressive of the legend, and heard diegetically by the characters as part of their conscious experience of the house, mountain, and valley.

The scratching sound at the end of the titles is an important element of the opening sequence of Episode 1 that establishes connections between the three central characters, the legend, and the landscape they inhabit. The first shot is a slow tilt down to the river, the setting corresponding with the point of view from the bottom of the hill by the standing stone established in the opening titles. Roger enters the frame swimming through the water and out of shot, the camera remaining static, fixed on the light glistening in the water. This dissolves into a shot of the ceiling in Alison's bedroom, with the reflected sunlight from the river shimmering on the blank surface. Alison traces patterns on the ceiling with her hand from the bed, circling over two spots of reflected light that flicker on the plaster. The shimmering on the ceiling suggests a phantasmagorical, translucent layer with another layer behind the physical surface. The confined space of the attic above the ceiling is where Alison and Gwyn discover the plates depicting abstract aspects of the legend. The light traced by Alison's fingertips is a figure for an archaeological imagination, with concealed artifacts hidden, buried, and then excavated or "unearthed," brought in to the light.

Alison and Gwyn hear a noise in the ceiling, a scratching sound on the surface like the claws of rats or mice, "something trying to get out." It is the sound of sharp nails scoring the rough grain of bare wooden boards. The dialogue subtly alludes to broader themes: on seeing an animal trap near the attic door, Gwyn suggests that former occupants "must have had this trouble here before," alluding like the plates to previous generations and the expression of the myth in this specific environment. The scratching continues throughout the scene, an off-screen sound where the source is not visible or identified, achieving the unsettling effect of an unseen presence in the house. The sound stops abruptly when Gwyn forces open the attic trapdoor to locate the source and discovers the stack of plates amid the dust and cobwebs and the scent of meadowsweet rising from the river in the heat. The senses of smell and hearing combine to convey flowers and claws, two aspects of the shifting figure of Blodeuedd, emanating from the setting, breaking through layers and experienced through the senses.

Opening the attic to locate the noise and touching the top plate on the stack ignites the power inherent in the landscape. In the novel, whenever Gwyn touches the plate, he makes allusions to the cinema: "everything went muzzy—you know how at the pictures it sometimes goes out of focus on the screen and then comes back? It was like that: only when I could see straight again, it was different somehow. Something had changed."[20] In the series this change, as well as the connections between the three characters in this landscape, is registered through editing, camerawork, and sound. When Gwyn touches the plate there is a sudden cut and shift in location back to the riverbank exterior, with a rapid zoom through the hole in the standing stone that frames Roger's face, accompanied by the brief rise and fall of harp music. The crash zoom and the music work alongside the cut to draw out the correspondences between the three characters and the artifacts at this moment, and the release of latent energy stored in the landscape. The sense of simultaneity across time and space is heightened by the intercutting between Alison and Gwyn with the plates and Roger by the river and the Stone of Gronw. The plates behind the shimmering surface of the ceiling and the standing stone by the river express aspects of myth. In the novel, the moment of the discovery of the plates in the opening chapter is paralleled in the next with Roger among the flowers by the river, where suddenly something "flew by him, a blink of dark on the leaves. It was heavy, and fast, and struck hard. He felt the vibration through the rock, and he heard a scream."[21] The harp music in the series conveys the vibration experienced by Roger as he jumps up in response to the sounds of the mountain and the valley. There is a sense here of a haunted landscape. Place generates the haunting, acting through people who feel and hear the vibration.

The scratching sound emanating from the attic is reprised at various points throughout the series, and is strongly associated with Alison. In the closing scene of the first episode, Alison moves and talks in her sleep, and the scratching sound rises in pitch and volume off-screen. In Episode 2, a montage of rapid close-ups transforms Alison's face into an abstract decorative floral design, as the scratching sound is heard again, as well as a loud screech and the fluttering of wings. It is as if Alison is transformed into an owl, the "flower-face" of *The Mabinogion*,[22] reflecting the cruel aspect of Blodeuedd. The sounds appear at moments of heightened emotional intensity, sonically representing a projection of interior emotional states. The scratching sound is also heard over shots of the matchboard paneling peeling away to reveal the wall painting of the lady framed by flowers and claws, and this is accompanied by a cut to a brief shot of Alison shuddering on her bed. Strange scratching and rustling noises are heard behind the padlocked door of a ruined outbuilding, as if an animal is trapped inside: the derelict room contains Bertram's motorcycle. At the end of the series Alison, possessed, emits a scream like the screeching of an owl while a storm rages outside, the mise-en-scène reflecting the character's interiority. The off-screen sounds of scratching, screeching, and fluttering are manifest in the location as if emanating diegetically from an unseen presence, heard over shots of the house and exterior surroundings.

The sound of a motorcycle is also heard off-screen. In the third episode there is an exterior shot of the house at night. The film cuts from the house to a shot of an owl in the trees, and a wavering and shimmering light briefly illuminates the bird. There is another cut back to the house in darkness as a spot of light from a headlamp sweeps across the exterior. On the soundtrack the sound of a motorcycle engine passes the house, approaching in the distance and then receding, all off-screen. There is another cut back to the owl staring into the camera and the sound of a hoot, the eerie note integrated with the shrillness of a flute. The off-screen sound of the motorcycle passing the house, the lamplight sweeping the exterior, corresponds with the figure in the photographs, and Bertram from the previous generation woven into the "present" iteration or expression of the myth. The unseen motorcyclist is a kind of ghost, but there is a sense that the landscape, as Gwyn observes, is not haunted but the events of the legend are still happening, with temporal moments interweaving "in mythic time where everything is simultaneously present,"[23] folding and pleating like the landscape surface and archaeological strata described by Lively. As with Alison's ceiling, the combination of shimmering circles of light and sound conveys the "past" in the "present." The surface layer of the landscape is deceptive and unstable, interweaving different times, rendered through the sound of the motorcycle moving through the landscape. In a later episode, there is a static exterior shot of a road as a motorcycle passes across the frame, the sound approaching then receding. The patterning of editing and sound establishes a narrative motif characterized by repetition and variation, the noise of the motorcycle heard off-screen and on-screen, outside of the frame and in. This patterned combination of on- and off-screen sound is also a feature of the complex narrative structure of *Red Shift*. The photographs, the text of *The Mabinogion*, the plates of the dinner service, the wall painting, the archaeological artifacts, and the sounds of scratching, fluttering wings, and a motorcycle engine in *The Owl Service*, intersect in the landscape as intrinsic components that express multiple aspects of the myth that are simultaneously present. In *Red Shift* an archaeological artifact, the constellation of Orion, and the sound of a cry, perform a similar function, creating the illusion of apparent perspective and acting as a kind of communication that collapses linear time and space.

Red Shift

In *Red Shift* protagonists from the Roman occupation of Britain, the English Civil War, and from the 1970s, experience the landscape around Mow Cop, Barthomley, and Rudheath in Cheshire simultaneously. There is no sense of linear chronological distance between the characters. The narrative structure does not adhere to a linear understanding of time, where different historical episodes succeed one another, but instead to what Mark Fisher calls, in reference to the novel, "mythic time."[24] There is no definitive separation of past, present, and future. Rather, the "different historical moments [are] in some sense synchronous," like the perception of travelling sound waves and vibrations, and the "scrambling of time" is "playing out some out-of-time pattern."[25] There is no "now." Like in *The Owl Service*, the "illusion of linearity [is] shattered by [the] eerie repetition and simultaneities of mythic time."[26] In this case the characters express variations of the ballad of Tam Lin and Janet. However, this is not explicitly telegraphed in the novel or the television adaptation in the same way as the figures from *The Mabinogion* are in *The Owl Service*. Myth provides structure to the narrative and the actions of the characters, removing them from linear time and presenting "a series of traumatic clusters"[27] like stars. In *Red Shift*, the three male characters Tom (Stephen Petcher), Thomas (Charles Bolton), and Macey (Andrew Byatt) experience this "broken time" through seizures and a kind of visionary interior fragmentation where they see faces and hear sounds. Thomas is asked what he sees and hears when he is "badly." He replies: "Colours. Blues and whites. Sounds […]. Noises. Sounds. All sorts. A face—'afeared. And his hands—pressing. Way off. I know him. Seen him many times. I know all about him. Is it me?". Thomas sees a face and hands and hears sounds at moments of emotional distress. Macey sees two hands pressing at him, a long way off and then near. Tom hears the sound of a cry approaching then receding on the M6 motorway, shared moments of pain in a shared landscape. The axe head artifact that they hold in their hands also connects the three characters.

The *Red Shift* film was produced for the BBC *Play for Today* strand in 1978, adapted by Garner and directed by John Mackenzie. An earlier *Play for Today*, *Penda's Fen* (Alan Clarke, 1974), written by David Rudkin, is also "strongly landscape based," conceived with a kind of "archaeological imagination."[28] Rudkin suggests that the physical surface of the landscape can "become a translucent film," making visible the old villages, buildings, and roads beneath, "looking at the surface as though it were a filter through which you could just glimpse a previous layer."[29] This archaeological sense of a stratified translucent landscape is also a feature of the settings, inhabitants, and excavated artifacts in *Red Shift*. In the novel the stratified landscapes are experienced as translucent superimpositions, with buildings seen in moments of heightened emotional stress, such as the tower of Barthomley church seen by Macey[30] or the folly castle on Mow Cop seen by Thomas.[31] However, these are not visions of older buildings but structures yet to be built. Linear temporal distinctions between past, present, and future give way to a multidimensional vision of the landscape characterized by simultaneity and synchronicity. In the adaptation the characters not only see the landscape through translucent strata, realized through editing, but they also hear sounds, simultaneously faraway and close.

The opening title sequence of *Red Shift* begins with clusters of white and red lights like stars floating across the screen, the white appearing to approach and the red recede. A series of dissolves moves from this pattern of lights to a shot of a rotating spiral nebula, and superimposed over this are a succession of still images of faces: first Tom, then a slow dissolve to Thomas, and then Macey. The dissolves and superimpositions suggest the complex conceptualization of time and the layering of the landscape, echoing the analogy of the stars in the night sky in "Inner Time." Garner's work can "speed out to the edges of the universe like Jodrell Bank looking for signals from afar," while being simultaneously grounded back through stratified geological and archaeological layers of time and the generations of "people who have walked this landscape."[32] The next dissolve in the titles returns to the clustered circles of white and red light and the shot is pulled into focus to reveal traffic on the M6 at night, making the connection between the stars, the physical landscape, and the people that inhabit it. Tom, Thomas, and Macey are generationally connected to the landscape that the motorway traverses beneath the stars. The title sequence closes with a series of point-of-view shots from the perspective of a moving vehicle on the motorway, approaching Tom and Jan walking by the side of the road, the camera panning toward the couple as it passes, and then receding from them.

The layered landscape and the correspondences between generations who have occupied the site are conveyed through the editing and sound. The opening scenes of the film focus on Tom and Jan as their relationship reaches an emotional crisis point. Jan is moving away, and they argue with Tom's parents in the caravan at Rudheath. As Tom is arguing and shouting, his words become jumbled and incoherent, as repressed emotions come to the surface. He leans forward with hunched shoulders, pressing his head against the glass windowpane as the pain rises to the surface. There is a sudden cut from a low-angled close-up of Tom's face pressed against the glass to a close-up of Macey in distress, his mouth agape, emitting short, rhythmic, cries of panic. Macey's eyeline appears to be facing and reacting to Tom as they experience heightened emotional strain in the same landscape. The abrupt cut and the sound of a cry express the correspondence between them. The brief shot of Macey's panic is followed by another cut back to an extreme close-up of Tom's face, the camera moving closer to his eyes and brow toward the center of his pain. There is then a further cut to a close-up of Tom's hands pushing through the glass of the caravan window that shatters into sharp splinters as his head follows through. The film cuts back to a close-up of the screaming Macey, who reels back as if retreating from the severity of Tom's discharged energy. The camera pulls back to reveal the interior of a tent under attack, the canvas being violently torn by spears from the outside. Macey is surrounded by pointed spears forced through the canvas, visually echoing the shards of glass from the window, and trapping Macey who continues to scream in panic.

The axe head, blood, and hands connect Tom, Macey, and Thomas. The editing style makes these connections weaving and pleating between the characters by splicing together matching images of the axe and bloodied hands. Macey uses the axe as a weapon, the violence triggered by interior emotional distress. His blood-soaked hand holds the long wooden handle and the polished stone head. There is a cut back to Tom and Jan through the broken glass of the caravan window, forming a frame of sharp splintered shards. Tom is washing the blood from his hands. The editing cuts back and forth between Tom and Macey with blood on their hands following outbursts of interior energy. The editing, connecting Tom and Macey through the details of blood and hands, is followed by a cut to Thomas's hands pulling the axe head out of the soil. There is no handle, the wood disintegrated after prolonged internment, and the circular hole for the handle is caked in mud. A close-up introduces Thomas's face looking down at the object in his hands, before cutting again to his hands wiping the mud from the polished surface of the artifact. Phil Ryan's electronic score carries across these shots, functioning to provide continuity, and a sense of parallel moments between generations. The structure of the narrative is fragmented and nonlinear, as Macey is later seen burying the object. The patterning of elements of style, the shot composition, editing, sound, and music here function to convey a meeting point, rupturing chronological, linear distance.

The locations connect the experiences of Tom and Jan (Lesley Dunlop), Thomas and Madge (Myra Frances), and Macey and the Girl (Veronica Quilligan) together. Tom and Jan cycle across the landscape, passing by the reflecting bowl of the radio telescope at Jodrell Bank, making connections between the stratified landscape and superimposed layers of time, and the "whispers" detected by the telescope. The editing fragments linear duration, cutting from Tom and Jan on

Mow Cop to Thomas on the roof of the steeple at Barthomley church gazing toward Mow Cop, or from Thomas to Tom and Jan in the churchyard below. Editing is used to convey simultaneity and co-presence, blurring linear temporal and spatial boundaries. Cutting between the protagonists that occupy the landscape functions in a similar way to the sound design, situating the locations outside of linear time, as stratified translucent spaces with porous boundaries.

Similar to *The Owl Service*, the sound of a motorcycle approaching and receding is used in *Red Shift* to convey a sense of nonlinear time and co-presence in the landscape. In a scene in which Tom and Jan, soon to be separated, are standing outside of the latter's house at night, a faint blue-silver light illuminates their faces as if from the moon and stars. The camera remains fixed on Tom and Jan, as the sound of an approaching motorcycle is heard on the soundtrack, speeding along the nearby M6 motorway, first approaching and then receding. The distant off-screen sound rises in volume and pitch as it draws closer, and four short bursts like a cry mingle with the sound of the engine as it passes. The sound of the motorcycle is defamiliarized, the layered noise recalling Macey's cries of panic. Tom looks off-screen responding to the direction of the sound. He moves away, and a close-up shows his head leaning forward and pressed against his clenched fists, recalling the position against the caravan window and the sights and sounds of the visions Thomas and Macey describe. Tom suggests to Jan that they "need a communication satellite," glancing toward the constellations of stars in the night sky, that remain off-screen. The camera stays fixed, in the same way as the sound of the passing motorcycle and cry was not accompanied by a cutaway reverse shot revealing the source. Tom nods toward Orion's Belt, identifying the constellation as a communication satellite: at 10 o'clock every night they will look at the stars and be together once every 24 hours. In the novel, Tom suggests that there's never "now," and Orion may not exist: "It's so far away, we're looking at it as it was when the Romans were here."[33] Tom makes the arrangement with the departing Jan, but it reflects his link with Macey. Orion binds Tom to Jan, but also to Macey, equating the cries of generations with the traveling "whispers" from stellar objects.

This temporal delay, distance, or gap between the perception of the stars in the 1970s and Roman times is conveyed through active off-screen sound. Through "apparent perspective" the distance is gone, like the whirr and click of the camera on a delayed setting and the appearance of the figures in the photographs in *The Owl Service*. The source of the sound from the M6 motorway is not shown but plays an active role in the narrative. The off-screen sound, as if from a motorcycle approaching, adjacent, then receding is layered over with a mysterious unexplained sound recognizable as Macey's cry on the same landscape at Rudheath. Tom looks around in the direction of the noise, responding to Macey's pain. The sudden, intrusive sound passing by, rising then falling, might suggest a haunted landscape. However, it is a more complex multidimensional conceptualization of time and space. The sound emanates from the landscape, outside of the frame, from the diegetic space of the scene, binding off-screen and on-screen space, but the source is indeterminate and ambiguous, making connections between characters and landscapes in apparently different temporal moments. Sound articulates the experience of the landscape, reflecting the psychological state of the protagonists, and conceptualizing the shared space across generations, the sense of simultaneity and contemporaneousness.

This chapter has explored the relationship between landscape and sound in specific sequences from the television adaptations of *The Owl Service* and *Red Shift*. The soundtrack works alongside other elements of style such as editing to convey the complex representation of landscape, archaeology, consciousness, and time. The "whirr" and "click" of the delayed exposure time, or the principle of traveling sound waves detected by the antennae of a radio telescope, act as a figure for the temporally complex stratified landscape, seemingly haunted by eerie sounds. The sounds of scratching claws, flapping wings, screeching noises, or cries of distress from an apparent "past," weave into the soundscape of the "present." Sounds "play back" on a recurring loop like sonic ghosts. However, the landscape is not so much haunted, but a translucent stratified space where generations of inhabitants are coexistent and simultaneously present.

Notes

1   Penelope Lively, *The Presence of the Past: An Introduction to Landscape History* (London: Collins, 1976), 9.
2   Ibid., 10.
3   Garner, "Inner Time," in *The Voice That Thunders: Essays and Lectures* (London: The Harvill Press, 1997), 113.
4   Ibid., 114.
5   Ibid.
6   Ibid.
7   Ibid., 106.

8   Ibid., 112.
9   Ibid., 115.
10 Alan Garner, "Flint," in *Cornerstones: Subterranean Writings*, ed. Mark Smalley (Dorset: Little Toller Books, 2018), 146.
11 Ibid., 149.
12 Ibid., 150.
13 Ibid.
14 Alan Garner, in *The Bronze Age Man of Jodrell Bank*, aired Wednesday May 21, 2014, on BBC Radio 4.
15 Gwyn Jones and Thomas Jones, introduction to *The Mabinogion* (London: J.M. Dent, 1949), ix.
16 Alan Garner, "The Beauty Things," in *The Voice That Thunders: Essays and Lectures* (London: The Harvill Press, 1997), 202.
17 Ibid., 205.
18 Stephen McKay, "*The Owl Service*: The Legend Unravelled," in *Alan Garner's* The Owl Service *Programme Notes* (London: Network DVD Booklet, 2008), 9.
19 Ibid., 13.
20 Alan Garner, *The Owl Service* (London: William Collins, 1967), 13–14.
21 Ibid., 11.
22 *The Mabinogion*, trans. Gwyn Jones and Thomas Jones (London: J.M. Dent, 1949), 74.
23 Garner, "The Beauty Things," 205.
24 Mark Fisher, *The Weird and the Eerie* (London: Repeater Books, 2016), 90.
25 Ibid., 93.
26 Ibid., 95.
27 Ibid., 96.
28 David Rudkin, "The Edge Is Where the Centre Is: David Rudkin and *Penda's Fen*," in *The Edge Is Where the Centre Is: David Rudkin and* Penda's Fen: *A Conversation*, ed. Gareth Evans, William Fowler and Sukhdev Sandhu (New York: Texte und Töne, 2014), 14–15.
29 Ibid., 15–16.
30 Garner, *Red Shift* (London: William Collins, 1973), 60.
31 Ibid., 122.
32 Garner, *Bronze Age Man*.
33 Garner, *Red Shift*, 39.

Concrète Spaces: Musique Concrète in Gus Van Sant's *Paranoid Park* (2007)

Jessica Shine

Reflecting on Gus Van Sant's films *Gerry* (2003), *Elephant* (2004), and *Last Days* (2005), the director's long-term sound-designer Leslie Shatz observed, "[y]ou have to get into the totality of the experience and not just the dialogue."[1] Shatz's comment expresses something fundamental about the experimental approach to cinema and to soundscapes undertaken by Van Sant in these three films, unofficially known as the "Death Trilogy." Van Sant himself denied the idea that the trilogy was "planned that way." However, he did concede that while all three films "were pretty much made independently," they all had "death and youth similarities."[2] In the same interview, however, he implied that there was a concerted effort to make something new with these films and stated that "there may be another film on the way, so it could change."[3] This fourth film was clearly *Paranoid Park* (2007), produced after the term "Death Trilogy" had been coined, but stylistically and thematically similar enough to extend the trilogy to a "Death Quartet" of sorts. Despite Van Sant's protestation to the contrary, there were many reasons for the emergence of the "Death Trilogy" moniker: the films were clearly bound together by similar visual aesthetics (long uninterrupted takes inspired by the films of Hungarian director Béla Tarr, cinematographer Harris Savides's signature "tracking-the-boy-from-behind Steadicam," and wide expansive images of clouds and skyscapes). The films were also thematically similar to each other in their focus on infamous deaths and they maintained a similar sonic throughout. A fundamental element to this sonic aesthetic was the consistent use of *musique concrète*.

Throughout the films in the "Death Quartet" there is a consistent blurring of boundaries between the traditionally delineated sonic spaces of diegetic and non-diegetic. *Musique concrète* haunts the diegetic space of the "Death Quartet," frequently seeming a part of the diegetic space initially only to be realized as external moments later when sounds that do not quite fit with the onscreen action materialize in the sonic space. Similarly, in *Paranoid Park* both the diegetic and non-diegetic sonic spaces are enmeshed together so much that one could easily be mistaken for the other; in several scenes throughout the film, music and sound point us toward another realm, away from the image. This chapter will focus on the role pre-existing *musique concrète* and electro-acoustic music play in shaping *Paranoid Park*, contextualizing it within the broader framework of the "Death Quartet" and with particular reference to *Elephant*, a film with which *Paranoid Park* shares many of the same sonic elements. Further to this it investigates the role of *musique concrete* in relation to sonic haunting, exploring its impact on the narrative of *Paranoid Park*, our impression of its protagonist, and our relationship with the audio-visual experience itself.

Concrète Spaces

Musique concrète permeates the "Death Quartet" soundscapes. Its use is central to narrative progression in each film and provides aesthetic uniformity to the films. In this regard, *musique concrète* is as important a stylistic component of the quartet as the camerawork, the young male protagonists, and the acting styles that bind the four films closely together. The narrative implications of *musique concrète* vary from film to film, from creating an oppressive and intrusive space for Alex (Alex Frost) in *Elephant*, to a daunting and unknowable place for Alex (Gabe Nevins) in *Paranoid Park*, to more desirable spaces for Blake (Michael Pitt)[4] in *Last Days*. However, it is almost always used in tandem with other sonic and visual elements to create barriers that challenge the traditional notion that music provides insight into a character's interior world. These films challenge the idea of music providing a film with a coherent or conventional sonic space, frequently problematizing the hierarchies of diegetic and non-diegetic sound, enmeshing one in the other, blurring and erasing these artificial boundaries.

There has been considerable debate amongst film music scholars about the usefulness of distinctions between diegetic and non-diegetic music. In her book *Hearing Film*, Anahid Kassabian argues that the arbitrary division between diegetic and non-diegetic music "describes a 'film' prior to the music, that constructs its narratively implied world silently."[5] She contends that this is inappropriate because the time and space of a film are constructed as much by its visual as its sonic elements, and that the artificial distinction between diegetic and non-diegetic "obscures music's role in producing the diegesis itself."[6] She further argues that the distinction creates a dichotomy that limits our understanding of film as an audiovisual medium and which "grossly"[7] reduces the function of film music.[8] Her most pertinent argument for this book chapter is her assertion that the dichotomy "cannot comfortably describe music that seems to fall 'in between' these categories, much less account for its different character" and that "[p]erhaps more importantly, it shifts critical attention away from features of the music—through its ability to match cues with the visual track—that coincide with the different possible narrative statuses."[9] Kassabian's points about music creating the diegesis itself resonate strongly with the way in which *musique concrète* is used in the "Death Quartet" as the music frequently becomes not only the sonic texture behind the scenes but the ambient foundation of the diegesis.

Robynn Stilwell also challenges the idea that music can be easily categorized into diegetic and non-diegetic spaces and instead argues that the soundtrack can be "a place of destabilization and ambiguity" that "highlights a gap in our understanding."[10] She argues that traditionally the ideas of non-diegetic and diegetic have been viewed as "adjacent bubbles" between which there is little transfer.[11] She and Jim Buhler, therefore, termed this threshold-like space as the "fantastical gap" arguing that this phrase captured both its "magic and its danger."[12] She writes:

> Fantastical can literally mean fantasy (cinematically, a musical number, dream, or flashback), and in fact this is one implication of the change of state ... but it can also mean, musically, an improvisation, the free play of possibility.[13]

The "fantastical gap" allows for applications of film music that thoroughly invert traditional notions of how music should work onscreen. Van Sant seems to take advantage of this freedom and potential, evidenced in his use of *musique concrète*. In *Paranoid Park* in particular, the "adjacent bubbles" often overlap, the sounds from each space enmeshing with the other with the result that our attention is drawn away from where it should be toward sounds and music that seem out of place, as if they were specters calling to us from another sonic realm, making us aware of the narrator's external presence from within the diegetic space.

In both *Elephant* and *Paranoid Park*, *musique concrète* falls between the categories of diegetic and non-diegetic, and though not located within the scene, acts as an indicator of narrative. It does this by being at once a part of the narrative space and often seeming to emanate from the film's diegeses and simultaneously seeming strange and noticeable and forcing the viewer to realize the fiction of the narrative. Ben Winters argues that film music is both "a sign of the fictional state of the world created on" and "an indicator that the universe in which the events we are watching takes place is not real" and therefore doesn't need to be assigned to a "separate level of narrative."[14] In the "Death Quartet" *musique concrète* concurrently emanates from the narrative space and also resists the viewer's desire to penetrate that same narrative space and become involved in the film.[15] While the *musique concrète* is located outside the diegesis, it is not instantly recognizable as non-diegetic music, nor does it function as a character theme to tell us more about the protagonists. *Musique concrète* augments the film's soundtrack, but also plays a wider role in establishing a connection between spaces in the films. Used in tandem with other sonic and visual elements to create spaces that often overlap, pre-existing *musique concrète* is spliced and edited with the diegetic sounds, making them difficult to separate.

The external space created by the *musique concrète* is enmeshed with the internal spaces of diegetic sounds fusing the two together. In *Elephant*, for example, as the camera tracks Nathan through the school, Hildegard Westerkamp's *Doors of Perception* (1989), which plays alongside Beethoven's *Sonata No. 14, in C Sharp Minor, Op. 27, No. 2*, "Moonlight Sonata" (n.d.), merges with the expected diegetic sounds of Nathan opening and closing doors. However, these door sounds are not diegetic at all and are actually sounds from Westerkamp's piece. The choir sounds that also originate from Westerkamp's piece, too, could be mistaken for diegetic sounds of a school-choir. Yet despite these moments of synchronicity, Westerkamp's piece on the whole seems *detached* from the image, while remaining a part of it. Kevin Donnelly argues that the fusion of Westerkamp's piece with the onscreen action at this particular moment in *Elephant* "sets up a sense of dislocation through the dichotomy of representational images and unrelated sounds as a replacement for those that we might imagine as 'belonging' to the images."[16] This synthesis does not always provide answers for the viewer, in fact, as the spaces collide, the films deviate further from traditional character-based notions of film narrative. In both *Elephant* and *Paranoid Park* space becomes an entity of its own. The space is not always visual; often it is acoustic. Equally, the space is not always "real"; it is as frequently imagined. These spaces are not always visible or tangible, but the viewer is always aware that there is something more than what we are being shown; there is something unexplained beneath the veil in both these films. *Musique concrète* is, in tandem with the cinematography and the performances of the actors, the creative agent for these spaces.

Paranoid Park

Musique concrète is integrated into the soundscape of *Paranoid Park* in much the same way as in *Elephant*. Frances White's composition *Walk through Resonant Landscape No.2* (1992), first used in *Elephant*, is also incorporated into the soundtrack here. The overall effect however, is overwhelmingly different. In *Paranoid Park*, much of the electroacoustic music reflects Alex's lethargic demeanor. The music has a dreamlike quality to it and, in tandem with the cinematography and Gabe Nevins's boyish, unblemished appearance, adds to the perception of Alex as innocent. Alex is no cold-blooded killer like his namesake from *Elephant*. Although he covers up his part in the killing by discarding his clothes and his skateboard, the film presents him as sympathetic rather than callous. Alex's friend Macy instructs him to write down what happened to him. She also adds the caveat that he has to write it to a friend, presumably to her. This makes Alex's tone far more open than it would be if he were talking to an adult. It is arguable then, that the soundscape reflects this openness but, by extension, it must also mirror the fact that Alex burns the letter in preference to sending it. In other words, though Alex frees himself of the secret by burning the letter, the secret effectively remains untold and hidden. The music must therefore reflect both the weight that is lifted from his conscience, but also that the secret remains kept, though we the audience are privy to it.

When we first meet Alex he is curled up in a chair writing the words "Paranoid Park" in his journal. In the background we can hear the meditative whirring of Ethan Rose's electroacoustic piece, "Song One" (2006). Alex's uncle Tommy passes by and he watches him nervously. Tommy returns to the kitchen where all the sounds he makes, including the rattling of bottles in the fridge, are unrealistically augmented, as if to signal that Alex is on edge and discomfited by his uncle's presence. The electroacoustic music conflates with the hyperreal sounds; the music is indicative of something peaceful, but the augmented sound effects, however, point toward uneasiness and it is not clear if either element of the soundtrack represents Alex's interior landscape. The camera jumps to a brief shot of Alex walking through dunes before cutting to the first skate sequence with "Song One" still playing. This sequence draws attention to the fact that we are not listening to location sounds of the film, but to a soundscape that acoustically mirrors Alex's feelings toward the park and we only get very short glimpses of the "real" vibes and the sounds that might attract him to the park in the first place. He ineloquently names off the kind of people who inhabit the park, but we never really get to see them as they are, except in the hazy slow-motion Super 8 footage that seems to reflect Alex's romanticized vision of the park as a refuge. Aside from a brief moment at the beginning of the skate sequence we don't get to hear very much of the hustle of the park. The music has a far more meditative and otherworldly vibe than typically more aggressive and immediate skate punk music. It is more attuned to Alex's feelings toward the park than with the park itself; this is how Alex views the park as an escape from the real world, a place to relax and to lose oneself in a crowd. It seems as if the park itself draws Alex out of his troubled life and allows him the freedom to just be in a moment, to just watch.

The choice of *musique concrète* functions to make these scenes an exercise in listening and aids the visual, providing an initial impression that this is Alex's point of audition. *Paranoid Park* does not present an objective sonic perspective of the skate park itself. The diegetic mythology of Paranoid Park as a place of both danger and escape clearly frames his point of audition and the insight we get into the park is sonically filtered through Alex's teenage, dreamy reminiscence of the park before it was spoiled for him by the death of the security guard. In this way, the mobilization of electroacoustic music in *Paranoid Park* works very differently to the use of *musique concrète* in *Elephant* where we are given no interior access at all. *Paranoid Park*, on the other hand, provides very little in the way of impartial sonic spaces.

Though "Song One" dominates the opening sequence, there are moments where Frances White's *Walk through Resonant Landscape No.2*[17] intersects and merges with Rose's piece. One such moment occurs just as we are drawn out of the skate sequence to a tracking shot of Alex walking through the dunes. Initially, one might assume that the wind sounds are diegetic, but then we hear the familiar bird calls of White's piece (assuming one has seen *Elephant*). Rose's piece is absent from the scene for a full minute, replaced totally by *Walk through Resonant Landscape No.2*. This minute coincides with the moment when Alex walks to the sea-side bench to begin writing his journal entries. This makes the piece seem diegetic as White's music can easily be mistaken for sound design, thus placing it in the realm between diegetic and non-diegetic sounds. Again, however, the realization that music is not, in fact, diegetic points us to a space outside of the scene, forcing us to question the origin of the sounds. White's music, then, acts like a spectral presence, haunting this scene with or without our knowledge.

This section of White's piece, with its foreboding tones, acts as an omen for the terrible things to come. As Alex begins to remember the park, *Walk through Resonant Landscape No.2* fades away and is replaced by diegetic sounds, only for these then to be replaced by the re-emergence of "Song One." "Song One" becomes the anchor piece for this scene, but again it is infiltrated by another piece of *musique concrète*, Robert Normandeu's *La Chambre Blanche* (1985–6). French female voices are inserted (Normandeu's piece also uses French speech but the voices we hear do not appear to be those from his piece). *Musique concrète* punctuates "Song One" with sounds that do not belong to it and disrupts its peaceful feel. This could be emblematic of the "paranoid" element of the park's name, or it could represent the fact that the park is now tainted for Alex. At this stage of the film, though, we are not sure what it is that happened to him. Once more, *musique concrète* pierces the calmness of the cinematography (and, in this instance, the calmness of the other soundtrack elements). Considering its apparent lack of musical structure, the *musique concrète* could simply be reflective of Alex's disordered and ineloquent narrative. However, much of the film refuses to didactically explain Alex's thoughts and feelings, and, as a consequence, such a reading of the music would not be fitting. Instead, we are given merely fleeting moments of access to his mental and emotional state that remain unexplained and intangible, at once making us feel close to him but also distant: the soundscape representing the gossamer, spectral nature of Alex's interior space.

Musique concrète dominates the film's soundtrack at the precise moments of Alex's withdrawal from the world around him. This is emphasized by the fact that the location sounds, which are overridden by the music, are heightened before the music's appearance, and made prominent once more when the music disappears. These sounds are noticeably louder than other diegetic sounds in the film as apparent in the scene with the rattling bottles cited earlier. These moments suggest careful manipulation of diegetic sounds that subtly reveals Alex's emotions, in an exemplary use of what Michel Chion calls "rendered sound." According to Chion,

> The film spectator recognizes sounds to be truthful, effective and fitting not so much if they reproduce what would be heard in the same situation in reality, but if they render (convey, express) the feelings associated with the situation.[18]

In the moments before the introduction of *musique concrète* we get some truth about Alex's state of mind, but it is deliberately subtle, seemingly operating on precisely the subconscious level that Chion discusses.

An excellent example of this technique occurs in the scene in which Jared (Jake Miller) suggests the boys head to the park. This scene is shown twice. In the first instance Alex's narrative is very audible, but in the second the dialogue between the boys is muted and competes with the sounds of the skateboards. A few moments later the scene cuts to the boys sitting in the park. Before Jared exits from the frame it is very clear that they are in a skate park, the boys' voices competing with the sounds of that environment. Here the soundtrack conforms to one of Annette Davison's suggestions for a "soundtrack-as-critique" where she argues "the organization of sonic elements is not strictly hierarchical" as it is in the conventional organization of film sounds in which dialogue usually supersedes all other elements of the soundtrack. In films that are critiquing conventional practice, Davison notes that instead, "other elements of the soundtrack are privileged over dialogue."[19] Here, Van Sant's film refuses the traditional exposition found in dialogue in favor of the grating and chaotic sounds of the skate park with the characters quite literally robbed of their voices. Van Sant's use of sound here is noticeable because it is unexpected; the traditional hierarchies of film sound, which viewers are accustomed to, are undercut and subverted. When Jared leaves, all these sounds and noises are replaced by "Song One" and *musique concrète*, which also goes against the traditionally held hierarchies that Davison ranks as dialogue followed by sound effects then music.[20] In Davison's argument she points to Godard's use of a similar technique, which, she argues, noticeably contravenes traditional Hollywood practice and therefore critiques it, or at least is in a dialogue with the conventions.[21] When the scene is replayed Alex is writing in his journal, remembering a conversation. It is the sound of skateboards and not the dialogue that bridges the present moment of the journal writing and the past conversation. The second time the scene plays it cuts to the newscast of the suspected murder; once again it is the exaggerated sounds of skateboards on asphalt rather than dialogue that act as a bridge between these cuts. The sudden cut to the newsreader suggests that even the sounds of skateboards are now distressing for Alex.

Throughout *Paranoid Park* heightened location sounds reflect the stress or emotional strain felt by Alex during the scene.[22] This further links *Paranoid Park* to the acoustic style of the other "Death Quartet" films in which heightened diegetic sounds feature throughout for narrative and aesthetic reasons. For example, in *Elephant* when John is trying to control his drunken father, the sounds of the car, its doors, engine, gears, etc. are all noticeably louder than expected, which reflects the awkward tension on screen. David Sonnenschein argues that what he calls concrete sounds (diegetic sounds) "might also fall into the category of musical sounds if they become dissociated from the diegesis and turn into a kind of sensorial or emotional element independent from the characters' space-time reality within the story."[23] This sensorial assault of sound occurs in *Paranoid Park*, where we can clearly hear sounds that are much louder than expected, and this loudness draws attention to the emotional effect of the sounds. For example, in the scene where Alex and the other skater boys are called to a group briefing by Detective Liu, the sound of Jared's skateboard on the hall floor is so loud that it almost drowns out the dialogue between the boys. In a similar fashion, the sounds that Alex makes, as he roots through drawers to find a plastic bag in which to put his clothes and scurries through the house trying to avoid detection by potentially suspicious neighbors, are unusually heightened. The sounds of *Paranoid Park* itself, though, are some of the most aggressive and noticeable, the sounds of the skate wheels rendered to be loud and grating and the background noise of the park itself cavernously swallowing up the dialogue of the characters as their voices compete with the cacophony of shouting voices and boards for sonic dominance. This correlates with the information that we receive from the skater boys who tell us that the Park is a terrifying and dangerous place.

However, the clamor and the noise of *Paranoid Park* that are later presented to us are entirely absent in this first foray into the Park. Instead, Frances White's music stands in for the diegetic location sounds, as if to indicate that the Park is not yet terrifying for Alex and that it has become a sanctuary of sorts for him, away from the melee of his home-life. Obviously, though, the Park does become a site of trauma for Alex and the sounds subsequently reflect this. After the traumatic incident, even Alex's home itself becomes a source of augmented and oppressive sound; Frances White's piece no longer plays instead of the diegetic sounds, but collides with and interlaces with them.

In the shower scene, in which Alex attempts to wash away evidence of manslaughter, the diegetic sounds of the shower become intertwined in the *musique concrète* itself and are almost indecipherable from Frances White's work. In contrast to the former scene, diegetic sounds become enmeshed with the pre-existing music in this scene. Kulezic-Wilson describes the significance of the audio/visual *musique concrète* in this scene, acknowledging the audio/visual synchronicity between the birds painted on the shower wall and the bird sounds in White's piece. She writes:

> bearing in mind the previous employment of the same track in *Elephant*, it becomes obvious that White's *Resonant Landscape* motivated a reversed process of audiovisual design in which sound determined the scene's visual content prior to its shooting. And in the best tradition of a Murchian metaphoric use of sound, the crescendo of the sound of water in this shower scene, punctuated by birdcalls, becomes the single moment in the whole film that suggests the depth of the fear and desperation tormenting a protagonist overpowered by spiritual/emotional inarticulateness and numbness.[24]

This shower scene is the first time that we are sure that we are lingering in the same acoustic sphere as the protagonist. Sound is an oppressive force in this scene, even dominating the visual elements. As a piercing drone grows ever louder, the picture darkens to such an extent that it is hard to decipher Alex's features. The edges of the frame are vignetted and out of focus, as if the eye itself cannot focus properly. As the music fades and we are left with nothing but a high-pitched drone, the light almost totally fades and Alex's face is completely obscured. Perhaps this is an attempt to emulate Alex's own covered eyes, but I suggest that it can be read as a visual enactment of how disorientating the moment is. This is made even more explicit by the unsettling sounds. The fragmented images of Alex in *Paranoid Park* may keep the viewer at a distance visually, but the soundscape reveals the depths of his fears; it is the unsettling sound images rather than the actual visual images that reveal his uncertainties. The shower scene in *Paranoid Park* provides the most obvious instance of these sound images. Though there are images of birds on the shower wall, it is the *sounds* of the birds fluttering that evokes their image, or at the very least makes their presence on the shower wall noticeable.

There is no doubt that Alex is suffering in this scene; he is clearly aware of what he has done. His fears are magnified by the music, which becomes the acoustic embodiment of his previously vocalized litany of worries on the Eastside Bridge. Though he disposes of his clothes and his skateboard, it is not out of callousness but rather out of panic. This moment contrasts significantly to the manner in which White's music is used in *Elephant*, where the characters' fears and motivations are never explained. In *Paranoid Park*, the music gestures toward an illumination of his interiority, but the camera refuses to penetrate it.

In an apparent allusion to the canteen scene in *Elephant*, in which Alex covers his head as the sounds of the canteen swell, Alex also covers his face from view in this scene. Here, though, the effect is markedly different. Evidently, the trauma that Alex feels at this particular moment is internalized, and by covering his eyes he is not blocking an external image but blocking the viewer's capacity to see the trauma on his face. Our visual points of identification are obscured. In contrast, we clearly hear Alex's trauma as the sounds of the shower reach piercing heights and the chirping and fluttering of the birds punctuates the drone, but it is not clear whether these sounds are audible to Alex or whether they emanate from him at all. Instead, their presence is an unsettling sonic assault on the viewer that quite literally overwhelms the image. As Alex covers his face and as the image dims, the sound increases in both intensity and volume. By obscuring the image, Van Sant intensifies the effect of the scene, a gesture that aligns the viewer's position with that of the protagonist. In *Paranoid Park* the overpowering sound effects do not indicate that the protagonist is overwhelmed, but the combination of the piercing sounds, sporadic fluttering of the birds wings, and the obscured visuals have a clear effect on both the viewer and the image, providing momentary insight into the trepidation felt by Alex. In *Elephant* on the other hand, we are never brought to the same point by either the soundscape or the image because the *musique concrète* compounds our inability to understand the motives of the killers.

Though the *musique concrète* gives us some sense of Alex's frame of mind, it does not always help us to understand precisely what he is feeling. The viewer is often left with no clear insights into the character in much the same way as we are blocked from understanding Alex in *Elephant*. Although the emotional blockades in *Elephant* reveal nothing about Alex's motivations or feelings, by extension they reveal a considerable amount about the impossibility of ever understanding his reasons for the massacre. *Paranoid Park* seems, on the other hand, to present a barrage of emotion, which is realized sonically. There is such a sustained sonic assault on the senses that it becomes almost impossible to decipher its meaning especially when we are provided with very few visual reference points for these emotions.

The *musique concrète* works in tandem with Christopher Doyle's cinematography to maintain a barrier between Alex and the viewer. Megan Ratner reflects upon this idea in her article "Paranoid Park: The Home Front," highlighting the connection between cinematography and the soundscape:

> In *Paranoid Park*, Doyle and sound designer Leslie Schatz [sic] maintain a tension between what Alex does (very little, actually, though he's often on the move) and his mental state as reflected in the soundscape and the camera's subtle off-angles … the best example is the shower scene, which Van Sant

doubles back on later in the film ... a slowed-down image of drops running off his boyishly shaggy hair. Alex seems to actually be submerged, as if he were drowning by drips.[25]

Unlike in *Elephant*, where the faces of Alex and Eric (the perpetrators of the high school shooting on which the film focuses) are rarely examined by the camera, Alex from *Paranoid Park* is almost interrogated by it. Though his *tabula rasa* expression makes him inscrutable, his boyish and innocent features lend a purity to these close-ups, the shower scene being the exception. Though the camera refuses to intrude into Alex's interior, the *musique concrète* unleashes it, or at the very least makes his emotions audible. Scout Tayfora argues that "music takes over when words cannot be found." However, that music is not always revelatory or insightful.[26] Though many of the other scenes show the camera scrutinizing Alex's face, the music is often so unexpected for the situation, or absent entirely and thus gives no real signifier for his emotions. Alex rarely looks directly at us. Instead he often appears to be looking somewhere off-screen, into the distance as it were, or closing his eyes. Through the apparatus of the camera the viewer scrutinizes Alex but he never returns the gaze.

Despite the association in the film of Rose's pieces with skating, one of them also appears in a traumatic scene during Detective Liu's talk with the skater kids. Liu hands out photos of the train attendant's corpse to the kids and Alex seems to become overwhelmed by the images. "Song Three" (2006) fills the soundscape as the camera changes to slow motion. Alex looks to the detective, who appears to gaze directly at him. Alex raises his hand as if to suggest that he will confess. The calmness of the music implies that perhaps he will finally get the weight off his chest. The shot cuts to Alex writing in his journal and then quickly cuts to a shot of Alex vomiting in the bathrooms. The scene then changes to a close-up shot of Alex's face surrounded by a blurred, dream-like background. Alex then declares "I had tried to put this part out of my mind, but Liu's picture brought it all back." As Alex continues to talk the camera transitions dreamily back in time to a shot of Scratch's face. As "Song Three" continues over Alex's narrative we are guided to the moment where Alex and Scratch decide to hitch a ride on the train. "Song Three" seems perfect for this moment, encapsulating the freedom of the open-air train ride and the innocence of Alex's face as he leans back and lets the wind flow through his hair. However, we know how this train ride ends and so does Alex. Unlike the shower scene, this music does not hint at Alex's feelings about this moment. Is it then just indicative of his inability to articulate in any cogent way his feelings about the event? Or is it just an aesthetic choice that fits well with the moment? The soundscape here occupies an ambiguous space somewhere between being inside and outside the moment. It is its own strange space that we can hear but not quite investigate.

Conclusion

In *Paranoid Park*, the soundtrack functions to erode and/or subvert preconceived notions about what stories should say and how they should be told. This subversion links *Paranoid Park* aesthetically and philosophically with the other films in the "Death Quartet," and especially with *Elephant* where the rumbling tones of the music add to the distancing effects of the camera-work and the boys' actions. *Elephant* deliberately embodies the disengagement that was perceived to be one of the many causes behind the Columbine massacre. The music accentuates the gap between the characters and their actions, which creates an overall sense of detachment. However, in *Paranoid Park*, the same effect distances us from a sympathetic character whose trauma he cannot truly articulate and we cannot fully understand. Not only does the use of *musique concrète* challenge our notions of traditional score, it allows these films to challenge our pre-conceived notions of how these stories should be told. Instead of giving us concrete, tangible insights into the character we are only given fleeting, ethereal moments of insight in musical and sonic modes that are often incomprehensible or misleading. Just as the film's main protagonist Alex is haunted by his experiences in the park, our experience of his story is equally haunted by the manner of its telling. We are constantly misdirected by sounds that lead us away from Alex's perspective to another space that seemingly has no meaning for the story. His own recollections as presented to the viewer are themselves haunted by spectral sounds such as screams, clinking glasses, fluttering wings, and all manner of other sounds that do not fit with either his recollection or the onscreen action.

These films prompt questions and refuse to supply any answers; instead, the films encourage active viewing and listening. This active participation is often a result of loud sounds, music, or screams being placed over seemingly innocuous visuals. This chimes with Walter Murch's statement:

> If the audience members can be brought to a point where they will bridge with their own imagination such an extreme distance between picture and sound, they will be rewarded with a correspondingly greater dimensionality of experience.[27]

In both the soundscape and the visual space Van Sant asks the viewer to make up their own minds, to use their imaginations to seek answers that ultimately may not be found.

Notes

1. Klinger, Gabe, "Interview with Leslie Shatz: Sound Auteur," *Undercurrent* 1, no.10 (2006).
2. Amy Taubin, "Blurred Exit," *Sight and Sound* 15, no.9 (2005): 16.
3. Taubin, "Blurred Exit," 16.
4. Blake is modeled on troubled Rockstar Kurt Cobain, and the film is an interpretation of Cobain's final days before his suicide.
5. Anahid Kassabian, *Hearing Film: Tracking Identifications in Contemporary Hollywood Film Music* (2002), 42.
6. Ibid., 43.
7. Ibid.
8. Ibid.
9. Ibid., 43.
10. Robynn J. Stilwell, "The Fantastical Gap between Diegetic and Nondiegetic" in *Beyond the Soundtrack: Representing Music in Cinema* (2007), ed. Daniel Goldmark, Lawrence Kramer and Richard D. Leppert, 186.
11. Ibid., 186.
12. Ibid., 187.
13. Ibid., 187.
14. Ben Winters, "The Non-Diegetic Fallacy: Film, Music, and Narrative Space," *Music & Letters* 91, no.2 (2010): 229.
15. See Winters' concept of intra-diegetic music for further analysis of this phenomenon in "The Non-Diegetic Fallacy".
16. K. J. Donnelly, *Occult Aesthetics: Synchronization in Sound Film* (Oxford: Oxford University Press, 2013), 86.
17. Also used in *Elephant*.
18. Michel Chion, *Audio-Vision: Sound on Screen* (New York: Columbia University Press, 1994), 109.
19. Annette Davison, *Hollywood Theory, Non-Hollywood Practice: Cinema Soundtracks in the 1980s and 1990s* (Farnham, Burlington: Ashgate, 2004), 196.
20. Ibid., 196.
21. Ibid., 81.

22 Isabella van Elferen notes that David Lynch too "turns up the volume on seemingly meaningless diegetic sounds" but the effect is more focused on creating a unique cinematic style than heightening or reflecting an emotional response to the scenes. Isabella Van Elferen, "Dream Timbre: Notes on Lynchian Sound Design," in *Music, Sound and Filmmakers: Sonic Style in Cinema* (New York: Routledge, 2012), 175–89.
23 David Sonnenschein, *Sound Design: The Expressive Power of Music, Voice, and Sound Effects in Cinema* (Los Angeles: Michael Wiese Productions, 2001), 20.
24 Danijela Kulezic-Wilson, "Sound Design Is the New Score," *Music, Sound, and the Moving Image* 2, no.2 (2008): 128.
25 Megan Ratner, "Paranoid Park: The Home Front," *Film Quarterly* 62, no.1 (2008): 19.
26 Scout Tayfora, "Video Essay. Anaphora: Gus Van Sant's *Paranoid Park*," *Mubi Notebook Video*, https://mubi.com/notebook/posts/video-essay-anaphora-gus-van-sant-s-paranoid-park, (accessed October 27, 2022).
27 Walter Murch, "Stretching Sound to Help the Mind See," *New York Times* 1 (2000): 21 (accessed May 2, 2022).

8

Haunted by Extinction: Sounding an Arctic Uncanny

Lisa Coulthard

In his influential book *The Soundscape: Our Sonic Environment and the Tuning of the World*, R. Murray Schafer comments on disappearing sounds in acoustic environments. The fact that certain sounds such as leather saddle-bags or school hand bells might someday be extinct motivates acoustic ecology research studies such as The World Soundscape Project, BBC's Save Our Sounds, and environmentally focused collections such as the Western Soundscape Archive. Alongside these archival projects, there are also considerations of sonic extinction in the rendered, constructed sounds of cinema and television. From the rattling of milk cans on horse-drawn vehicles to the disappearing sounds of a natural world redolent with diverse animal life, cinematic soundscapes meticulously recreate a multitude of sounds that no longer exist in everyday life. This resonance of extinct, forgotten sounds is crucial to soundscapes of period films that mine such details for authenticity, impact, and historical detail. They also exploit such sounds for sonic nostalgia. In addition to sonically rendering archaic and historical objects, cinema also reincarnates the sonically extinct sounds of acoustic ecologies, landscapes, and environmental sounds. Acoustic ecologies of extinct animal and natural life push the sonic nostalgia of endangered sonic objects toward an uncanny haunting. In cinematic soundscapes of extinct animal and natural life, we hear the ghosts of the landscape itself. The eerie ecological ghosts of reconstructed endangered and disappeared acoustic ecologies allow the environmental dead to come back to life; they haunt the present by sounding extinction.

These uncanny landscapes of the sonic dead offer potential for an ecological critique of environmental destruction. For instance, if we consider the ecological hauntings of endangered sonic natural histories in Terence Malick's *The New*

World (2005), we note that bird calls dominate the soundtrack of the 1607 Virginian landscape. This acoustic aviary presence is striking in its variety, plenitude, and volume. However, the sound team was challenged by the tragic histories of species extinction: namely the disappearance of the Carolina Parrot and the Passenger Pigeon, two species that would have flourished in this area in 1607 but which were extinct by the early twentieth century. Of these, it was projected that the Carolina Parrot would have sonically dominated the landscape. However, there are no recordings of what this bird would have sounded like. Malick and his crew, in consultation with bird song experts at Cornell University's Lab of Ornithology, designed a sound that closely approximated Carolina parrot song through a complex analysis of body size, beak shape, and sub-species characteristics. Haunted by this sonically reanimated dead, *The New World*'s soundtrack is both an uncanny reminder of loss and a site for ecological critiques about extinction and colonization.

Turning these considerations of sonic extinctions further north, this chapter considers two series acoustically haunted by the ecological catastrophe of lost polar landscapes and disappearing Arctic ice. In what follows I analyze the uncanny sonic geographies of extinction in two AMC series focused on tales of Arctic death in the mid-nineteenth century: *The Terror: Season 1* (2018) and *The North Water* (2021). Drawing on Katherine Bowers's (2017) consideration of the polar Gothic (a literary mode that frames Nordic and Arctic exploration in terms of liminal spaces, supernaturalism, and haunting), I analyze the spectrality of Arctic soundscapes in which "ice creates a negative space, which gives rise to supernatural beings that reflect the self."[1] This negative space of ice is further haunted by the actual disappearance of ice in this current moment of climate catastrophe. Ice, as a sonic category, haunts both historical documents of the nineteenth century as well as current films and series about this era. As Bowers and others have noted, the Arctic imaginary is sonic, characterized by silence, echoed winds, enormously noisy ice cracks and groans, and the sounds of animal life both above and below water. Focusing on these two series of nineteenth-century Arctic shipwrecks, this chapter argues that more than resurrecting extinct sounds, sound in both *The Terror* and *The North Water* haunts these period landscapes with an awareness of their own pastness, with a recognition of contemporary Arctic ice's disappearance. The popularity of these series in an age of Arctic climate catastrophe is no coincidence. Rather than being merely nostalgic, the sonic spectrality of these series instead highlights what is already gone, never to be found again.

Sounding Solastagia: Uncanny Ecologies and the Spectrality of North

In a piece entitled "Uncanny Landscape" Jean Luc Nancy notes the rise of landscape painting is coincident with a secular turn; where God was once there to relate man to nature, now there is nothingness. It is this nothingness that Nancy contends defines landscape painting. The secular relation of human to land becomes one of estrangement, dislocation, and invisibility. Landscape announces itself as a space in which the human figure is "unsettled, straying and lost";[2] it is absence as presence, a contradiction that accounts for its fundamental uncanniness. This absence as presence is particularly notable in the conceptual landscape of "North." The idea of North is an abstraction, a direction, a blank slate of isolated white space of ice and snow not associated with large human populations or industrial noise. Not a geography or a place, North is a direction and, although it is tied to terms such as the Arctic or the polar, it is also a relative idea as North "moves always out of reach."[3] As Sherrill Grace notes, North is "multiple, shifting and elastic,"[4] its coordinates always overshadowed by ideological, political, and aesthetic orientations. Framed as mysterious, haunted by danger and disaster, North is also, as Barraclough, Cudmore and Donecker point out, characterized by supernatural and metaphysical forces: "mysterious occurrences, otherworldly beings, and sorcerous inhabitants."[5] Shane McCorristine characterizes this "spectral Arctic" as one imbricated with the histories of polar expeditions, exploration, and colonialism. As he notes, ideas of the Arctic "drew on actual experiences and cultural imaginings of dreams and other supernatural phenomena in the far North."[6] Focusing on the spectres of the Franklin expedition, he addresses the fundamental uncanniness that was always a part of far North colonial exploration: "spectral experiences such as dreaming, clairvoyante travel, reverie, spiritualism and ghost-seeing informed ideas of the Arctic and the searches for a Northwest Passage through the Arctic."[7] Katherine Bowers draws on these associations in her framing of the genre "polar Gothic," which posits North as an "extreme and uncanny space at the end of the world."[8] Most accounts of a spectral or polar Gothic are founded on the idea that North was experienced by colonizer-explorers as supernatural and otherworldly; they also note that our current spectral Arctic is haunted by this history of traumatic, failed expeditions. As well, these hauntings highlight an additional phantom layer—the ghostly victims of deadly Arctic and polar colonialism and imperialism. "North" in the twenty-first century is haunted by all of these spectres as well as by current and future ecological catastrophe and destruction: it is haunted by its own destruction and death.

The ghosts that return to haunt us in an era of climate catastrophe are laden with past lives and concerns for future destruction and disappearance. In his work on hauntologies, Mark Fisher comments that it's not the past but "the spectres of lost futures" that haunt us.[9] Nowhere is this more present than in the ecology of climate change and crisis. It is not merely that we are haunted by futures that failed to happen, it is also the fact that we are living in an era where the future itself is being mourned as a loss. As Gan et al note, "Ghosts remind us that we live in an impossible present—a time of rupture, a world haunted with the threat of extinction."[10] Landscape is haunted by material pasts and traces, but also by futures of extinction and destruction.

While Fisher notes that the future "is always experienced as a haunting,"[11] the eerie landscapes of extinction point to a more specific haunted absence. It is not merely the "no longer" and "not yet" outlined by Fisher, but the radical "will never be again" of extinction that possesses twenty-first-century Arctic landscapes. Turning from the idea of the North more generally to the specificity of the Arctic highlights the importance of ice in these hauntings. As Timothy Morton notes, "Aesthetic images of the environment are always predicated on disaster,"[12] and this is widely evident in the icy Arctic landscape haunted by present extinction. The presence of polar ice has always been considered uncanny, eerie, and mysterious and its disappearance is even more so. As an "indicator region,"[13] the Arctic is at the centre of today's environmental crisis and melting ice "has become an iconic symbol of the Anthropocene."[14] Summarizing what Mark Serreze calls the Arctic "death spiral,"[15] Peter Wadhams states without exaggeration that "the loss of Arctic ice is a threat to us all"[16] and "an unmitigated disaster for the earth."[17]

That this death spiral is central to our current climate catastrophe and the accompanying states of mind associated with ecological distress cannot be overstated. Glenn Albrecht et al have developed the term "solastalgia" to describe the sense of desolation for absent or lost solace usually found through positive connections to landscape, environment, homeland, and space. More specifically, they tie this concept to a sense of "environmentally induced distress," of which ecological catastrophe is a significant driver.[18] The loss of Arctic ice is a nexus for this distress as it represents a tipping point and concretization of the catastrophic landscape and world-changing effects of the climate crisis human beings have created. It is also arguably central because of the uncanny imaginaries historically associated with North: the conceptualization of North as pure and purifying, uncanny, mysterious, and supernatural intensifies the horror, melancholy, and distress of its disappearance. What was already uncanny becomes truly horrifying when it is haunted by its own disappearance, a catastrophe created by colonial and imperial commercialism, which are in turn intertwined with the anthropocenic destruction of the planet. If as Gan et al. argue, that every landscape is "haunted by past ways of life,"[19] the Arctic is shadowed by an additional haunting of its disappearance. The Arctic is not merely another landscape haunted by "erased histories and voices buried"[20] (although it is also this); it is a landscape of an absent future, a geography of extinction.

The disappearing ice of the Arctic landscape connects solastalgia to haunted extinctions, which is in turn related to the sonic extinction with which I began. The idea of North has long been a sonically inflected one. Analyzing Samuel Taylor Coleridge's 1798's "Rime of the Ancient Mariner," Bowers notes the otherworldly nature of polar Gothic's soundscapes: it's a "land of ice, and of fearful sounds"; the ice "cracked and growled, and roared and howled"; the otherworldly sounds are "like noises in a swound."[21] Relying on "destabilization and disorientation through aural and visual distortion,"[22] the polar Gothic's sonic conventions include "creaking ship sounds, an eerie, muffled silence."[23] For many, the Arctic is an imagined soundscape of eerie silence and soothing, meditative white noise. From eighteenth-century explorers' accounts of the supernatural sounds of ice and wind, to Robert Payne's 1970's album *Songs of the Humpback Whale*, to the 12-hour soothing white noise track of an Arctic icebreaker that became popular in 2017, the North's uncanny spectrality is sonically defined. It is no surprise then that the absent future of a disappeared Arctic is equally tied to sound and music. In what follows, I analyze the uncanny solastalgia that haunts the melancholic soundtracks of two Arctic themed series highlighting the historical horrors and supernaturalism of eighteenth-century Arctic exploration: *The North Water* and *The Terror: Season 1*.

Sonic Extinction: Listening to the Hauntings of Arctic Noir

Connected to the violent landscapes of colonial exploration and exploitation, the historically inflected Arctics of *The Terror: Season 1* and *The North Water* foreground polar Gothicism, supernaturalism, and the uncanny. Both are narratives of circa 1850 Arctic ship crews stranded on ice and fated to die. The similarity of these series has been widely noted.[24] Although more overtly present in *The Terror*, both series are haunted by the Franklin expedition. *The Terror*, an adaptation of Dan Simmons's novel of the same name, is a reimagining of the Franklin expedition. In this account, however, the culprit is not the cold or tainted rations. It is instead a hybrid polar bear/human beast, which is called the Tuunbaq by the local indigenous people. After an indigenous shaman is shot and killed by an explorer and his daughter Silna is taken onto the ship, the Tuunbaq begins to hunt and kill Franklin, his crew, and officers. Stuck in ice, running out of food, suffering from lead poisoning from tainted canned goods, and finding themselves hunted by the Tuunbaq, the majority of the crew devolves into states of insanity, hallucination, and cannibalism. Except for the fictional character Crozier, everyone dies. Most are killed by the beast, some by murder or suicide. While the Tuunbaq is obviously a departure from the conventional, historical Franklin narrative, the series makes it unclear whether (given the crews' poisoning, starvation, and trauma) the Tuunbaq exists at all. Instead, it animates what McCorristine would argue has always been present in Arctic exploration narratives—the uncanny, the spectral, the supernatural.

Eschewing the supernatural in favor of a more human form of evil, *The North Water*—based on Ian McGuire's novel of the same name—addresses a nineteenth-century whaling ship's encounters with disaster and death. The plot follows two lines of horror: the planned destruction of the ship by its owner for financial gain, and the murderous impulses of the animalistic evil of whaler Henry Drax (Colin Farrell). The focalizer for the narrative is a disgraced and laudanum addicted surgeon Patrick Sumner (Jack O'Connell), who joins the ship after his traumatic imperialist service in India. During his military service as surgeon, he participated (at the behest of his superiors) in looting during the Siege of Delhi, an event portrayed in flashback during the series. Ambushed in the act of thieving, Sumner gets brutally shot in the leg and watches as his fellow officers are killed. He then witnesses the murder of an Indian child by a British guard. His wounding in these events as well his scapegoating by those higher in command and eventual discharge is what has led to him to his current state addiction and desolation. On the whaling ship, Sumner discovers the rape of cabin boy Joseph Hannah (Stephen McMillan). The likely culprit is Drax, a beast of a man with no ethics, conscience, or social ties. As the story continues, Drax murders Hannah because he is afraid of being outed as the rapist, which is the first in a series of murders. Alongside this plot is a

Franklin-related narrative of a ship stuck in ice, stranded men, and certain death. Unlike *The Terror* though, the fate of the ship and crew is tied to brutal and avaristic capitalism and colonizing imperialism, not supernaturalism. That the ship owner is pursuing an insurance scam highlights the fact that the whaling industry at this time was already dying because of overhunting. After the ship sinks and the rescue ship is blown apart in a storm, the survivors slowly die except for Drax and Sumner, who have a confrontation back in England in the final episode.

As noted, both series rely on the Franklin expedition as a spectral subtext that haunts their landscapes. As McCorristine argues "the Franklin expedition has come to occupy a spectral place in contemporary culture,"[25] an uncanniness notable in its connection to the supernaturalism of the Victorian era as well as in the twenty-first century's recognition of the roots of climate catastrophe in the commercial exploitation, colonial imperialism, and industrialization that the era of Arctic expeditions represents. As he notes apropos of Simmons's novel *The Terror*, the future appears as a haunting: "Simmons has this psychic couple [Crozier and Silna] foresee a future in which westerners will invade the Arctic causing the ice to melt and its people to degenerate."[26] The Franklin expedition also concretizes climate change when we note that the Northwest Passage is nearly now ice-free in the summer months.[27] Ian McGuire's novel *The North Water* is equally engaged with narratives of imperialist capitalism, colonial histories of expedition and conquest, and the extinctions wrought by climate catastrophe. In an essay for *The Times*, McGuire notes the disappearance of sea ice, which he contends was almost as important as whales for nineteenth-century whalers; as proof of this he cites Greenland whaler Scoresby's 100-page description of ice in his diaries.[28] He goes on to comment that although whaling once drove many economies and cities, by 1840, 200 years of overhunting decimated the population almost to the point of extinction, whaling had ceased, and the decline in Arctic ice began.

The North Water's production history echoes this narrative of glacial disappearance: the series' makers claim that it was shot further north than any other series.[29] Where *The Terror* used extensive CGI to create its north, *The North Water* was filmed on location, which proved a difficult task because they could not find sufficient ice until they hit 22 miles south of the north pole. This claim that it was shot so far north is not mere swagger. It is a narrative of changing climate that we have seen in other film production histories such as Alejandro González Iñárritu's *The Revenant* (2016) and Quentin Tarantino's *Django*

Unchained (2012). Twenty-first-century filming crews frequently struggle to find locations cold enough for icebound or snow laden locations.[30]

The differences in location shooting versus a CGI-created north in *The North Water* and *The Terror* respectively also impacted the soundscapes of the two series. Both series have what could be characterized as electro ambient scores composed by highly regarded contemporary experimental composers, musicians, and sound artists who do not primarily work in television or film, the recently deceased Marcus Fjellström (*The Terror*) and Tim Hecker (*The North Water*). Contemporary Arctic focused series have a tendency toward techno ambient electronic music: for instance, recent Nordic noir series have featured scores by Johann Johannsson, Hildur Guðnadóttir, Ben Frost, and Rutger Hoedemaekers. More than indicating a shared community of Canadian or Nordic musical kinship, these correlations suggest a certain aesthetic dominant and convention for Arctic scoring—one that highlights ambience, distortion, sonic disorientation through electronic manipulation of instruments, unrecognizable slowed down musical phrases, and phased white noise. It also points to the way that ambient, electronic music can meld into the sounds of the polar Gothic noted earlier, eerie silence, wind gusts, ice cracks, ship moans. We see this in the soundtracks of the series noted as well as in albums like Ugasanie and Dronny Darko's *Polar Gates*,[31] SleepResearch_Facility's *Deep Frieze*,[32] Vancouver's Loscil's "Goat Mountain,"[33] as well as David Bickley and Tom Green's *Erebus & Terror*.[34] It is no coincidence that Loscil and Green and Bickly have tracks that appear in *The Terror*.

As these examples indicate, ambient Arctic as an identifiable sonic genre aligns with the Arctic sublime and the polar Gothic. Words like chilling, unnerving, claustrophobic, dark, icy are repeatedly put in play to describe the soundtracks of *The Terror* and *The North Water*. More than just setting tone, atmosphere, and landscape, these icy scores engage in ecological critique through their haunting ambience. Not just echoing and articulating the Arctic sublime of eerie silence and screeching ice, this twenty-first-century Arctic ambient music is haunted by a disappearing and endangered landscape. Music and sound effects in these historical, period-focused series work to create a sense of extinct sounds both in the Schaefferian definition as sounds that are no longer with us and in the sense of the acoustic parameters of extinction itself. In the series I consider here, this ecologically inflected sonic haunting (extinct sounds and sounding extinction) highlights sounds related to land and seascapes: whale song, ice cracks, wind noise.

In *The North* Water, it is the Canadian electronic musician and composer Tim Hecker in combination with the Canadian sound team headed by supervising sound editor Jane Tattersall that highlights the sonic polar Gothic of extinction. Hecker's score in particular not only represents but also haunts the Arctic landscapes. As one reviewer notes, "Hecker's music manages to pull you into places you'd rather not be. Into the cold and wet, the dark and foul."[35] Dominated by reverberant wide-open spaces of ice and snow, with "sounds that crunch and crack like ice and whistle and swirl like biting winter wind" and are "as chilling as the icy waters seen on screen,"[36] the score was always envisioned as a scoring of landscape as well as character and theme: "The survivalism entailed and the bleakness of the wasteland was great tableau for music. When I was speaking with Andrew (Haigh) beforehand, he was clear about wanting the music to play a prominent role in that landscape."[37] Rising above these geo-immersive sounds however, is "the most broken hearted whale song."[38] Hecker notes that, in addition to listening to lengthy YouTube videos of ship sounds like the Arctic icebreaker, he listened to live Arctic whale song webcasts because he wanted "to mimic the sounds of a whale."[39] Nonetheless, he wanted these sounds to be abstracted, poetic, rather than literal, indexical, or metaphorical:

> I started with an examination of *glissandi* (note and pitch bending). I used techniques to pitch track a cello and its fundamental tone with a sine wave, so it would bend and almost become a sonar transfiguration. A lot of orchestral samples want you to stay within the twelve-tone system and there's so much more to it when you bend between notes. If I'm not writing for an instrumentalist in a studio, I'll contort samples so they'll be more flexible and expressive than they are given to you out of the box.[40]

If we consider episodes 2 "We Men Are Wretched Things" and 3 "Homo Homini Lupus" there are a few key moments where whale sounds appear that stress this metaphorical sonic presence. It should also be noted that this is not a singular object but takes forms variable in pitch, timbre, and tone, with the melancholic strain evident in each. In episode 2 ("We Men Are Wretched Things"), the first appearance of this whale song-like element is when Otto questions whether Sumner died on the ice. Quoting Swedenborg, Otto says to Sumner, "You would have met the dead, spoke with them." The second is after Drax leaves Sumner's cabin after getting his wound dressed and the motif sound bridges to a shot of a glacier. It then appears during the buildup to the first whale hunt, and it appears intermittently throughout the hunt. It also appears in a pitched form at the end of the autopsy scene of the cabin boy Joseph Hannah. In episode 3 ("Homo Homini Lupus") the first appearance is a sound bridge between the first scene (a discussion inside the ship about the intentional sinking of the ship betweens Captains Brownlee and Morwood) and the second (a confrontation on deck between Henry Drax and Patrick Sumner). In the first scene, whale song blended with the low-frequency wood creaks and ice scrapes along the sides of the ship, create an underlying tension to their scheming. As the discussion comes to a close, Brownlee reflects on Drax's crime of raping a cabin boy: "evil times we live in" he says as the whale song music comes up in the mix and bridges to the following scene that confirms the evil noted in the first. Similar whale sounds (although pitched differently) come up again in this second scene combined with ice cracking in a dialogue pause before Drax leans in and threatens Sumner with "I'd say a dead cabin boy is the least of our fucking problems." Sounds in the same whale-like register occur again when Drax is physically examined, when the cabin boy's tooth is found in his arm, and when Drax threatens Sumner with a stick.

Appearing intermittently throughout the series and becoming more clearly aligned with Sumner, the mournful whale song is not only metonymic for the whales who are slaughtered mercilessly and pointlessly (since the ship will be sinking anyway) or for their species which is being hunted to extinction. It is also a sonic motif tied to trauma and human violence. The whale cries are reflected in Brownlee's cries of pain as well as other scenes of human suffering, emotional as well as physical. So rather than whale song, these whale-like elements in the score focus on human alienation, trauma, violence, and brutality, of which the whales are also victims. The sound appears again later in episode 3 in the deliberate sinking of the ship, which itself takes on whale-like elements. Throughout these scenes, there are also the constants of wood creaks from a moving ship and ice sounds that serve not only to remind us of location and landscape but that work in concert with the more eerie elements of Hecker's buzzing phased score that creates a tone of doom. Similarly, the violent cracking of ice heard voluminously in scenes on deck and in muted form when within the ship haunts the episodes before the sinking of ships and the stranding of the men; the violent ice cracks are a function of the ships invading their space, a sound that mutes in the mix after the men are stranded on the ice itself and the ships are gone.

The sonic haunting in *The Terror* is more pervasive and less distinctly musical than that in *The North Water*. The composer Marcus Fjellstrom died while composing the score so it is unclear whether the sparse musical cues are a function of this or an aesthetic choice. Regardless, music is isolated to certain scenes and sound effects come to the fore. In contrast to *The North Water*'s on-location shooting and sound, *The Terror* was shot entirely on a sound stage and relies on CGI and post-production sound to create its Arctic world. As sound editor Lee Walpole comments, "It's not often that we get a chance to 'world-create' to that extent and in that fashion ... The sound isn't just there in the background supporting the story. Sound becomes a principal character of the show."[41] The sound was built entirely in post-production and Walpole worked hard to recreate authentic historical sounds. He spent time recording on a replica of Sir Francis Drake's ship Golden Hind, took kayaks out on the frozen Thames for water and ice sounds, and hydrophone recordings from a frozen lake in Canada. For snow, Walpole eschewed Foley, instead he "wanted to get an authentic snow creak and crunch, to have the character of the snow marry up with the depth and freshness of the snow we see at specific points in the story. Any movement from our characters out on the pack ice was track-laid, step-by-step, with live recordings in snow. No studio Foley feet were recorded at all."[42] Ice pings and crunches from the frozen lake and the frozen Thames were treated, pitched, and slowed down, and amplified to sound like great sheets of ice. For the sounds of ice crushing the ship, he used windmill wood sounds: "As the situation gets more dire, the sound becomes shorter and sharper, with close, squealing creaks that sound as though the cabins themselves are warping and being pulled apart."[43] This icy haunting even extends to the Tuunbaq itself which is a combination of treated human voice, recorded bear sounds, and dry ice: "I turned to dry ice screeches and worked those into the voice to bring a supernatural flavor and to tie the creature into the icy landscape that it comes from."[44]

A dominant sonic trope is the pervasiveness of Arctic wind in the series. Many sound editors note that wind sounds are difficult to characterize and must be constantly varied in order to not sound like mere noise or disturbance. In *The Terror* each wind sound has its own character and presence, a focus that allows them to attain a musical identity with distinct timbre, contour, tone, and pitch. As executive producer Soo Hugh comments, "Whenever we're in an exterior scene, we use different wind ... Through all 10 episodes, the wind is different."[45] As she goes on to note in the final episode where Tuunbaq is confronted and killed:

> In the Tuunbaq massacre scene, we cut the wind down dramatically during the attack. We wanted to create this interesting vacuum, the sense that these men

are trapped in this one moment and there is a sense of timelessness. There's the element of very, very light winter but it's almost oppressively thin. Then later on when Lady Silence comes back and she finds the dead Tuunbaq, the wind comes back in and we went for a mournful, musical tone.[46]

Wind is distinct from ice cracking or whale song; it is, however, equally characterized as an Arctic sound and the winds in *The Terror* stress isolation, a landscape without plant life (there are no leaf rustles for instance), open spaces with loud, whistling wind. As sound editor Walpole notes, "Outside, I wanted it to feel almost like an alien planet. I constructed a palette of designed wind beds for that purpose."[47] The effect is one of icy coldness. There is no doubt that the winds of *The Terror* are icy Nordic winds, impossible in an Arctic tundra with Spruce trees and without ice.[48] The winds haunt the characters, their fate, but also our contemporary sense of the Arctic haunted by the disappearance of this Arctic cold.

In the introduction to the book *Arts of Living on a Damaged Planet: Ghosts and Monsters of the Anthropocene*, Elaine Gan, Anna Tsing, Heather Swanson, and Nils Bubandt argue that the "winds of the Anthropocene carry ghosts,"[49] traces of the past, and the eerie future of a "world haunted with the threat of extinction."[50] As they note, winds "are hard to pin down, and yet material; they might convey some of our own sense of haunting."[51] Sound designers, editors, and mixers often comment on the problem of wind as a sound effect: as mere air, it has no actual sound; rather it makes sound when it contacts surfaces: leaves, grasses, buildings, ground, trees. For polar Gothic, the Arctic landscape requires an icy windscape—that is, wind sounds that only contact the brute rock faces and ice of an empty, desolate, Nordic landscape. Arctic winds are part of the polar Gothic's eerie and haunted silence. *The Terror* plays with this gothic uncanniness by downplaying musical score, which is ambient and played low in the mix, and foregrounding wind. Wind operates as an ambient context that reminds us of the open and empty landscapes of North and that creates a nostalgia for a lost world that is only palpable through the passing and ephemeral sounds of wind. Wind reminds us of the past tense of this narrative and the past of the Arctic itself. This Arctic wind is pitted against the encroaching violence of colonialism, imperialism, human-created extinction, and climate catastrophe.

Adding to these haunting landscapes of Arctic desolation, death, and climate catastrophe, both series also stress the hallucinatory, unreal sonic elements of Arctic landscapes that form a part of the polar Gothic that Bowers describes and the spectral Arctic articulated by McCorristine. In each of the novels as well as in the literature of exploration analyzed by Bowers and McCorristine, the sonic Gothic elements of Arctic landscapes dominate; ice screams, groans, cries out in almost human voices. In the novel *The Terror*, the ship is "always rumbling, moaning";[52] the ice's daytime moaning "turns to screams"[53] and at night, the "hull groans, the frigid deck moans under their feet."[54] McGuire's *The North Water* is similarly sonic: "the wind is too loud and all around the ice is screeching"; the "great percussions of the ice field, the thunderous explosion of one floe meeting another one";[55] "the noise outside is enormous."[56] Because the loudness of this screeching and thunderous ice would be too sonically dominating over the numerous hours of each series, the presence of ice cracks and scrapes tend to be pervasive rather than sudden claps of noise. The underlying noise of ice is there throughout the ship scenes in muted form rather than occupying a dominant place on each soundtrack. Music, wind noise, and contact noises of ice with ship dominate more than the aural landscape of ice itself. Yet, the underlying noise is ever present in the maritime scenes, adding a layer of tension, a sense of ice as a living, threatening thing, rather than just an empty landscape.

In addition to the screams of ice, the cold, darkness, and strange lights of the Arctic create an unreal landscape that reflects and infects the mindsets of the colonialists and whalers. North is not merely a landscape but a place that changes one's thinking, confuses one's bearings, muddies thoughts, and makes the lines between reality and fiction blur. Both series focus on sound to highlight these states of mind: in *The Terror* it is episode 15: "Terror Camp Clear" that most clearly engages these hallucinatory hauntings; in *The North Water*, it is Sumner's laudanum laced imaginings that frame his Arctic experience within a wider context of colonial terror. The sound team for *The North Water* relished the freedom that these hallucinatory scenes offered for playing with sound: Brownlee's screams of pain, ice creaks, and Drax's voice find their way into memories from colonial Delhi, music and sound effects become warped and distorted as a laudanum haze determines the soundscape. This same hallucinatory sonicity comes back in episode 4 "The Devils of the Earth" with the confrontation with the polar bear, which is not inflected by laudanum haze but rather by pure desperation, starvation, desolation, and the buzzy, mixed scores with traces of vocality effectively creates the frozen chill of this moment. Similarly, the toxic food and crazed lunacy wrought by starvation and Arctic dislocation in *The Terror* come to a climax in the fog-laden scenes of "Terror Camp Clear" that turn to wind sounds, beast roars, human screams, and a rumbling, distorted score to reflect the disorientation, confusion, and fear of that sequence.

From Arctic-inspired hallucinatory aurality to the Gothic wooden creak of ships compressed by ice to the wild icy winds, mournful whale song, and low-frequency hums of Arctic landscapes, the soundscapes of both series are haunted by the traces of a nineteenth-century spectral polar Gothic tied to supernaturalism. In addition to these hauntings that can be framed in terms of the concept of solastalgia noted earlier, there is a second strain of ecological critique, or at least ecohorrific lamentation in the sounds of whales hunted to the brink of extinction, their songs cut short by hunting and marine vessels that interfere with their sonic communication. This lamentation of extinctions extends beyond animal life to include ice creaks, breaks, and crashing of plates and wild winds across open spaces. All of these sonic markers of an Arctic landscape are endangered, if not yet extinct. Listening to these acoustic resurrections can intensify awareness of their loss and absence. The hauntings become doubled as the traces of the past become compounded by the failures of the future. The icy sonicity of ambient electronic music or field recordings distanced from themselves by pitching, filtering, and treating, as well as the low level hums of noise, wind, effects, and music in both series convey a historical Arctic, but also a disappearing one. And the two are imbricated as the rise of commercial exploration of North is directly tied to the future destruction of that same North; although these series could be linked with the nostalgic revival of whaling and labor songs of Shantytok, there is also the more reflexive and critical strain that makes clear that nineteenth-century colonialism, industrialization, and capitalism were the beginning of the end. This era is the tipping point of climate change, which creates another layer of haunting for the Arctic landscape. It also indicates the extent to which sound can perform effective ecological critique. The disappearing Arctic is evident not in the polar images CGI'd or filmed on location. It is felt in the sonic Arctic, which has always formed the core of the uncanny polar imaginary.

Notes

1   Katherine Bowers, "Haunted Ice, Fearful Sounds, and the Arctic Sublime: Exploring Nineteenth-Century Polar Gothic Space," *Gothic Studies* 19, no.2 (2017): 72.
2   Jean-Luc Nancy, "Uncanny Landscapes," in *The Ground of the Image* (New York: Fordham University Press, 2005), 57.
3   Peter Davidson, *The Idea of North* (London: Reaktion Books, 2005), 8.

4   Sherrill Grace, *Canada and the Idea of the North* (Montreal and Kingston: McGill-Queens University Press, 2001), 16.
5   Eleanor Rosamund Barraclough, Danielle Marie Cudmore and Stefan Donecker, "Introduction," in *Imagining the Supernatural North*, ed. Eleanor Rosamund Barraclough, Danielle Marie Cudmore and Stefan Donecker (Edmonton: University of Alberta Press, 2022), xiii.
6   Shane McCorristine, *The Spectral Arctic: A History of Dreams and Ghosts in Polar Exploration* (London: UCL Press, 2018), 3.
7   Ibid., 3–4.
8   Bowers, "Haunted Ice," 72.
9   Mark Fisher, *Ghosts of My Life: Writing on Depression, Hauntology and Lost Futures* (Winchester and Washington: Zero Books, 2014), 27.
10 Elaine Gan, Anna Lowenhaupt Tsing, Heather Anne Swanson and Nils Bubandt, "Introduction: Haunted Landscapes of the Anthropocene," in *Arts of Living on a Damaged Planet: Ghosts of the Anthropocene*, ed. Anna Lowenhaupt Tsing, Heather Anne Swanson, Elaine Gan and Nils Bubandt (Minneapolis: University of Minnesota Press, 2017), 6.
11 Mark Fisher, "What Is Hauntology?," *Film Quarterly* 66, no.1 (2012): 16.
12 Timothy Morton, "Romantic Disaster Ecology: Blake, Shelley, Wordsworth," *Romantic Circles*, University of Colorado Boulder, January 2012, https://romantic-circles.org/praxis/disaster/HTML/praxis.2012.morton.html (accessed August 25, 2022).
13 Sarah Jaquette Ray and Kevin Maier, "Introduction: Approaching Critical Northern Issues Critically," in *Critical Norths: Space, Nature, Theory*, ed. Sarah Jaquette Ray and Kevin Maier (Fairbanks: University of Alaska Press, 2017), 2.
14 Markku Lehtimäki, Arja Rosenholm and Vlad Strukov, "Introduction: Visualising the Arctic," in *Visual Representations of the Arctic: Imagining Shimmering Worlds in Culture, Literature and Politics*, ed. Markku Lehtimäki, Arja Rosenholm and Vlad Strukov (New York and London: Routledge, 2017), 6.
15 Shani Cairns, "The Arctic Death Spital," *Scientists' Warning Foundation*, December 2022, https://www.scientistswarning.org/2022/01/12/arctic-death-spiral/ (accessed August 25, 2022).
16 Peter Wadhams, *A Farewell to Ice: A Report from the Arctic* (New York: Oxford University Press, 2017), 4.
17 Ibid., 104.
18 Glenn Albrecht, Gina-Maree Sartore, Linda Connor, Nick Higginbotham, Sonia Freeman, Brian Kelly, Helen Stain, Anne Tonna and Georgia Pollard, "Solastalgia: The Distress Caused by Environmental Change," *Australas Psychiatry* 15, no.1 (2007): 95–8.
19 Gan et al, *Arts of Living*, 2.

20 Sladja Blazan, "Haunting and Nature: An Introduction," in *Haunted Nature: Entanglements of the Human and the Nonhuman* (Cham: Palgrave Macmillan, 2022), 9.
21 Bowers, "Haunted Ice," 74.
22 Ibid., 77.
23 Ibid., 76.
24 Bathsheba Demuth, "Arctic Horror Is Having a Comeback," *The Atlantic*, September 2021, https://www.theatlantic.com/science/archive/2021/09/the-terror-north-water-arctic-history/620153/.; Ben Lindbergh and Miles Surrey, "Between 'The North Water' and 'The Terror,' AMC Is Obsessed with Gloomy Nautical Dramas," theringer.com, The Ringer, August 2021, https://www.theringer.com/tv/2021/8/16/22623929/amc-the-terror-the-north-water-colin-farrell (accessed August 25, 2022).
25 McCorristine, *The Spectral Arctic*, 201.
26 Ibid., 215.
27 See "A Nearly Ice-Free Northwest Passage," *NASA Earth Observatory*, https://earthobservatory.nasa.gov/images/88597/a-nearly-ice-free-northwest-passage#:~:text=In%20mid%2DAugust%202016%2C%20the,breaks%20up%20to%20varying%20degrees (accessed August 25, 2022).
28 Ian McGuire, "Death on the Ice: The Real Story behind the Arctic Whaling Drama," *The Times*, September 2021.
29 Dalya Alberge, "Arctic Thriller's Film Crew Struggled to Find True Frozen Waste," *The Guardian*, August 2021, https://www.theguardian.com/tv-and-radio/2021/aug/01/arctic-thrillers-film-crew-struggled-to-find-true-frozen-waste (accessed August 25, 2022).
30 Nick Romano, "How Global Warming Stopped The Revenant From Filming Its Final Scene," *Cinema Blend*, July 2015, https://www.cinemablend.com/new/How-Global-Warming-Stopped-Revenant-From-Filming-Its-Final-Scene-72777.html (accessed August 25, 2022); "Wyoming Snow Attracts Quentin Tarantino for Django Unchained," *The Location Guide*, October 2012, https://www.thelocationguide.com/2012/10/wyoming-snow-attracts-quentin-tarantino-for-django-unchained/ (accessed August 25, 2022).
31 Michael Barnett, "Ugasanie & Dronny Darko—Arctic Gates—Review," *This Is Darkness*, February 2019, http://www.thisisdarkness.com/tag/polar-ambient/ (accessed August 25, 2022).
32 Sleep Research Facility, "Deep Frieze (CSR72CD)," *Cold Spring*, https://coldspring.bandcamp.com/album/deep-frieze-csr72cd (accessed August 25, 2022).
33 Loscil—Topic, "Goat Mountain," *YouTube*, August 2022, https://www.youtube.com/watch?v=F3UUvBeBs2w.

34 Bickley and Tom Green, "Erebus & Terror," *Another Fine Day*, https://anotherfineday.bandcamp.com/album/erebus-terror (accessed August 25, 2022).
35 Charles Steinberg, "Tim Hecker on His First Score for the North Water," *Composer*, accessed August 25, 2022, https://composer.spitfireaudio.com/en/articles/tim-hecker-on-his-first-score-for-the-north-water.
36 Bill Pearis, "Tim Hecker Brings Icy Dread to 'The North Water' Score (stream a track)," *Brooklyn Vegan*, September 2021, https://www.brooklynvegan.com/tim-hecker-brings-icy-dread-to-the-north-water-score-stream-a-track/ (accessed August 25, 2022).
37 Steinberg, "Tim Hecker".
38 Simon Tucker, "Tim Hecker—The North Water: Soundtrack Review," *At the Barrier*, October 2021, https://atthebarrier.com/2021/10/15/tim-hecker-the-north-water-soundtrack-review/ (accessed August 25, 2022).
39 Steinberg, "Tim Hecker," n.p.
40 Ibid., n.p.
41 Jennifer Walden, "Capturing, Creating Historical Sounds for AMC's *The Terror*," *PostPerspective*, April 2018, https://postperspective.com/capturing-creating-historical-sounds-amcs-terror/ (accessed August 25, 2022).
42 Ibid.
43 Ibid.
44 Ibid.
45 Steve Greene, "'The Terror': The Crafting of a Thrilling Ending That Found Dignity and Hope in One Final Swarm of Tragedy," *IndieWire*, May 2018, https://www.indiewire.com/2018/05/amc-the-terror-episode-10-ending-spoilers-finale-1201966788/ (accessed August 25, 2022).
46 Ibid.
47 Walden, "Capturing, Creating Historical Sounds".
48 Clarisa Diaz, "Spruce Trees Have Arrived in the Arctic Tundra a Century Ahead of Schedule," *Quartz*, August 2022, https://qz.com/spruce-trees-have-arrived-in-the-arctic-tundra-a-centur-1849406537 (accessed August 25, 2022).
49 Gan et al., *Arts of Living*, 1.
50 Ibid., 6.
51 Ibid., 4.
52 Dan Simmons, *The Terror* (New York: Little, Brown, and Co., 2007), 10.
53 Ibid., 36.
54 Ibid., 41.
55 Ian McGuire, *The North Water* (New York: Henry Holt and Company, 2016), 47.
56 Ibid., 155.

Acoustic Ghosts and Haunted Landscapes in Contemporary British Landscape Cinema

Aimee Mollaghan

In "Invention, Memory, and Place,"[1] Edward Said asserts that geography as a socially constructed concept can invest a geographical location with particular mythological significance and power through the creation of collective memory and narrative. Further to this, geographer Donald W. Meinig asserts that "every mature nation has its symbolic landscapes. They are part of the iconography of nationhood, part of the shared sets of ideas and memories and feelings which bind a people together."[2] The word "landscape" implies something that is composed or constructed, a shaping of the natural environment, something ideological that potentially informs the manner in which one might see or experience the world. Landscape is a text that is subject to multiple readings or interpretations. And, if landscape is essentially a text that can be read, it is also a text that can be written. In a similar fashion, film as a constructed text has the power to fashion cinematic landscapes through the creation of audiovisual narratives imposed on a geographic locale. Touching on theories gleaned from cultural geography, hauntology, acoustic ecology, and trauma studies, this chapter draws on these ideas in order to investigate the notion of fabricated cinematic narratives imposed on specific rural geographic locales through a process of sonic haunting in British historical dramas that have emerged since the turn of the Millennium.

These dramas, loose adaptations of classic European novels such as *Wuthering Heights* (Emily Brontë, 1847), *Lady Macbeth of the Mtsensk District* (Nikolai Leskov, 1865), *La Salle Terre* (Honoré de Balzac, 1799), and *Sunset Song* (Lewis Grassic Gibbon, 1932),[3] specifically locate themselves within the nineteenth- and early twentieth-century British rural landscape, irrespective of the original setting. They incorporate gothic or supernatural elements, often altering the rendering of supernatural aspects of the original narrative present in the source novel. The characters are not haunted by tangible ghosts in these films, but rather the landscape is haunted by ineffable revenants. Rather than using landscape as a physical canvas for the unfolding of action, films such as *Wuthering Heights* (Andrea Arnold, 2011) and *Sunset Song* (Terences Davies, 2015), the two case studies under consideration here, use representations of landscapes haunted by hyperreal soundscapes and acoustic ghosts in order to allow for a psychological engagement with these constructed environments.

Mythological Landscapes and Trauma

Said contends that if symbolic landscapes did not exist there would be an existential need to invent them in order to facilitate the creation of a collective memory that supports the development of a national identity.[4] I would suggest that part of this psychological restoration is the creation of mythological narratives, particularly in relation to landscapes. One finds that there is often an establishment or re-establishment of identities bound up in the landscape as a reaction against threat or trauma to the extent that it can be constructed or reconstructed as a mythical or utopian space. This is visible in Ireland in the nineteenth century with the establishment of associations such as *Conradh na Gaeilge* (the Gaelic League) arguably demonstrating nostalgia for a period before anglicization and positioning the west of Ireland as an idealized environment in which to establish Irish identity. This imposition of narratives and order on the landscape can also have geopolitical consequences. The rise of national socialism in Germany is concomitant with the promotion of a *Volk* identity linked to the expansion of Germany's homeland. A restorative narrative of land and nature associated with the national soul was woven in reaction to defeat and loss of territory in the wake of the First World War, this *Volk* narrative creating a physical and ideological tie with the soil mythologizes the countryside as the natural home of the "simple" German people.

Cinema is in a unique position to fashion mythological haunted landscapes through its ability to sculpt with image, sound, and time. The landscape in *Sunset Song* is not the actual Scottish landscape but a symbolic version of it, the verdant countryside of New Zealand where much of the film was shot, serving as an idealized nostalgic version of home. These fabricated landscapes, heavy with trauma, are often bound up with the weight of nostalgia and a desire to fashion a sense of collective or individual identity. Svetlana Boym defines nostalgia as "a longing for a home that no longer exists or has never existed."[5] Referring to nostalgia as an epidemic or disease, Boym claims that nostalgia reappears as a defense mechanism in the face of upheaval and upset, potentially serving as a psychological curative or helping to restore an individual sense of social connection.[6]

Further to this, Laurence Kirmayer avers that our efforts to process trauma or disruption results in the creation of a specific narrative landscape grounded in folk models of memory. These reconstructions of traumatic memory involve building narrative landscapes in order to contain these memories, present them to others in a convincing manner or to simply confront them. Kirmayer, recalling Said, points out that if a trauma is shared by an entire community, it creates a potential public space for retelling: "If a community agrees traumatic events occurred and interweaves this fact into its identity, then collective memory survives and individual memory can find a place (albeit transformed) within that landscape."[7]

Unlike the way in which the landscape is positioned as a restorative utopian space in contemporary Irish landscape film such as *Silence* (Pat Collins, 2012), the landscapes of the films under consideration here are not necessarily landscapes of regeneration, nor are they the well-behaved topographies of the Romantic painterly tradition or indeed, traditional British period dramas. These films are etched with narratives of violence, brutality, and degradation. These are unctuous, disembogued unheimlich andscapes of excess, scarred by psychological trauma.

In Andrea Arnold's loose adaption of *Wuthering Heights* Heathcliff (James Howson and Solomon Glave [young Heathcliff]) and Cathy (Kaya Scodelario and Shannon Beer [young Cathy]) might be the central characters but one cannot call them innocent or virtuous victims. Severed from his family, from his homeland, the racially ambiguous Heathcliff's frustrations or feelings are borne out in acts of unspeakable violence and cruelty to both his fellow humans and animals. In the world of this film, violence begets violence inscribing on the landscape a cyclical violence borne of intergenerational trauma. A scene where Heathcliff callously and wordlessly hangs his future wife Isabella's (Nichola Burley and Eve Coverley [young Isabella]) dog on a gate is echoed later in the film by his bullying foster brother Hareton's (Michael Hughes) foul-mouthed feral son, who impassively dangles a puppy on a gate. The film is replete with moments of violence such as these. Heathcliff is regularly beaten by members of the Earnshaw household and in return he regularly subjects his wife Isabella to brutal savagery. Cathy and Heathcliff express their fervent affection for each other with acts of violence. In a recurring scene, Cathy violently wrenches a tuft of Heathcliff's hair from his head and sends it off on the wind to haunt the moor. Even the seemingly ladylike Isabella is moved to express her desire for Heathcliff through violence, jealously attacking and drawing blood from her sister-in-law Cathy. Heathcliff, traumatized by Cathy's seeming rejection of him, her marriage to the landed Edgar Lynton (James Northcote and Jonny Powell [young Edgar]), and her subsequent death, attempts to transgress boundaries between the phenomenological and spectral realms. On the death of Cathy, he breaks into the gentile "big house" of the Lyntons and moves on top of her corpse, pulling her arms around him and stroking her hair. He later digs up her grave, attempting to open her coffin and hasten a reconciliation above the earth.

In a similar fashion, Terence Davies seems to have etched the Scottish landscape of *Sunset Song*, his 2015 adaption of Lewis Grassic Gibbon's eponymous literary classic, with many of the ghostly and psychological concerns explored in his 1988 film *Distant Voices, Still Lives*. Protagonist Chris Guthrie (Agnes Deyn) stoically and mutely endures the trauma of her mother's suicide following the infanticide of her twin baby siblings, and an existence of abuse at the hands of her brutal father and later by her husband Ewan (Kevin Guthrie), who is traumatized by the prospect of fighting in the Great War. Rather than merely examine these ideas through the narrative of the film, the cruelty and hardship experienced by the protagonists are embodied in the landscape, the driving rain, and ferine wind whipping through the Scottish countryside in pathetic fallacy. Chris, who is often considered to personify the figure of Caledonia, makes explicit the connection between her and the land. In a voiceover toward the end of *Sunset Song*, she professes the following:

> Everything was changing and as the land changed, so did Chris. She looked for the days gone by. She looked to see the faces of her mother and father in the firelight before their lamps were lit. Faces dear and close to her. She wanted to hear the words they'd known and used in the far off youngness of their lives; Scots words to tell to your heart how they rung and held it through all of the toil of their days and the unending fight with the land and a queer thought came to her, nothing endured but the land. The sea, sky and the folk who lived there were but a breath but the land endured and at that moment she felt in the gloaming that she was the land.[8]

Trace and Hauntology

Intertwined in this idea of mythological landscapes impregnated with unspeakable trauma is the notion of trace and hauntology. There are two predominant strains of Hauntology that are useful to my argument. Arguably, the best known is Jacques Derrida's Marxist derived play on the French homonym of ontologie in *Specters of Marx*,[9] which as Colin Davis asserts "supplants its near-homonym ontology, replacing the priority of being and presence with the figure of the ghost as that which is neither present nor absent, neither dead nor alive."[10] Although Derrida never adequately defines what he means by trace he regards it, not as a presence but rather as a postmodern simulacrum of a displaced manifestation that makes reference beyond itself. For the purposes of this chapter I am taking trace to be analogous to his notion of spectral trace and I wish to consider the natural environment within *Sunset Song* and *Wuthering Heights* in relation to the rendering of the landscape through hauntological or spectral presence, specifically through the medium of sound.

The second source of hauntology that is useful to my argument is associated with psychoanalysts Nicolas Abraham and Maria Torok. Echoing Karl Marx's assertion that "all the traditions of all the dead generations weigh like a nightmare on the brain of the living,"[11] they postulate the idea of a "transgenerational communication"[12] in which the concealed traumas of previous generations have the potential to disturb the lives of their descendants even and especially if they know nothing about their distant causes. Their concept of hauntology provides an alternative manner in which to consider ghosts as a way of encrypting the unspeakable secrets of past generations.

Abraham and Torok conceptualize their phantoms as a dead ancestor "still intent on preventing its traumatic and usually shameful secrets from coming to light."[13] Derrida's specters, conversely, are gesturing toward a future yet to be inscribed while Abraham and Torok's Phantoms actively lie about the past, willfully misleading the haunted subject in order to preserve their secrets. Derrida's specters are shady indistinct figures that exist between life and death, between past and present, between presence and absence. They watch, sight unseen, without reciprocity. They are, in the words of Derrida, "neither substance, nor essence, nor existence."[14] Derrida is not interested in restoring the ghost to the order of knowledge and encourages us to communicate with the specter, not with the expectation that they will reveal a secret to us but rather, to allow us to experience an "essential unknowing which underlies and may undermine what we think we know."[15] The secrets of Derrida's specters are not unspeakable because they are proscribed, but because we do not yet have the language to articulate them and yet cinema with its own peculiar grammar is a particularly useful medium for the presentation of trauma and nostalgia, providing a conduit for specters to whisper their secrets. In Ken McMullen's experimental film *Ghost Dance* (1983), Derrida infamously refers to cinema as "the art of ghosts, a battle of phantoms, … [or] the art of allowing ghosts to come back."[16] He also asserts, "I believe that ghosts are part of the future, the modern technology of images like cinema, enhances the power of ghosts and their ability to haunt us."[17] Further to this, he suggests that "the technology of images' affects" our behavior and perception of temporality.[18] Reflecting on Derrida's rumination on the temporality of spectrality, Akira Mizuta Lippit suggests that Derrida turns assumptions about ghosts returning from the past on their head and instead suggests that specters exist in the present, haunting the present in advance.[19] To paraphrase Derrida, haunting is historical but it is not dated. It does not have a specific chronology or time. It functions outside of temporality, outside of linear narrative. *Sunset Song* and *Wuthering Heights* are erstwhile period dramas and yet of no period. This is echoed by Mark Fisher who asserts, "[t]he future is always experienced as a haunting,"[20] as something that is always encroaching on the present.

Boym, ostensibly reiterating Fisher's framing of hauntology, as "a crisis of space as well as time,"[21] considers the origins of nostalgia to be historical rather than psychological, associating the spread of nostalgia with not only a sense of spatial dislocation but also a temporal one. Vietnam war veterans and Holocaust survivors tend to speak of traumatic events in the present when called on to provide witness testimony. Haunted by trauma, they experience events or time out of sequence. Jonathan Schell, reflecting on the temporality of the postwar era writes,

> In imagining any other event, we look ahead to a moment that is still within the stream of human time, which is to say within a time in which other human beings still exist, and will be responding to whatever they see, looking back to represent time and looking forward to future times that will themselves be within the sequence of human time. But in imaging extinction, we gaze past everything human to a dead time that falls outside the human tenses of past, present and future.[22]

This disjunction in time has resulted in new ways of internalizing past and future in the face of chronological progression and spatial expansion. Indeed, some of the symptoms associated with nostalgia as an ailment include a temporal dislocation between the past and present, a longing for a native land, confusion between the real and imaginary, and the ability to hear or see ghosts, symptoms that are spectrally manifested in the films under consideration here. Auditory nostalgia associated with the music and sounds of home, such as that experienced by characters in *Wuthering Heights* and *Sunset Song*, is particularly apparent in those suffering from nostalgia.

This modern nostalgia, whether connected to war or existential trauma, is "mourning for the impossibility of mythical return, for the loss of an enchanted world with clear borders and values."[23] This is arguably what is happening in both *Wuthering Heights* and *Sunset Song*. Both films are episodic, their chronologies off-kilter. *Wuthering Heights* is haunted by unseen phantoms, phantoms whose voices lurk out of time, crossing temporalities; Cathy's singing winging its way through the primeval winds of the Yorkshire moors, the ubiquitous sound of lapwings, the crying of dogs, the tap-tap-tap of branches on a bedroom window returning throughout the film to unsettle us and point us toward a time out of joint. Similarly, Fisher asserts that these sorts of "spectres are unsettling because they are that which cannot, by their very nature (or lack of nature), ever be fully seen; gaps in Being, they can only dwell at the periphery of the sensible, in glimmers, shimmers, suggestions."[24]

In *Echographies of Television*, Derrida presents the specter as a paradoxical entity, as both "visible and invisible, both phenomenal and nonphenomenal, as a trace that marks the present with its absence in advance."[25] In addition, Derrida writes that the specter cannot merely be considered to be a "visible invisible" that can be seen by the subject. It is someone who watches or concerns the subject without any reciprocity. As Lippit points out, Derrida draws on notions of spectral visuality to propose a paradoxicality of vision. Derrida's invisibility haunts the visible. Rather than an inversion of visuality it is another form of visuality, constituting, in Lippit's words, a "seeing blindness."[26] These are films where, spectral presence is marked by traces of what Lucretious might refer to as bodies "in motion that we see in sunbeams, moved by blows that remain invisible."[27] The landscapes are haunted, not necessarily by spectral visibility or invisibility of image, but rather by sound.

The ghosts or phantoms manifest themselves through their aurality. In *Sunset Song*, at the point where a traumatized Chris makes peace with Ewan's execution for cowardice and his previous egregious treatment of her, he haunts the family house in rural Scotland as a munificent specter offering Chris comfort and psychological regeneration. His ghostly presence is marked by shards of light radiating through the void of the parlor as his soft bodyless voice intones "I've come home." There is a similar ambiguity to the haunting in *Wuthering Heights*. We never see the revenant of the deceased Cathy, rather her spectral presence is experienced as though a fragment of memory; her childish song echoing across the moors, the recurring tap tapping of a branch on a window, her presence carried on the particles of debris transported by the wind which diffuse through the open window of wuthering heights to swirl around Heathcliff's body as he lies on the bedroom floor mourning her death.

Temporality and Sonic Haunting

Bearing this sonic haunting in mind it is, perhaps, not a stretch to consider the role of sound and music in the creation of the mythological landscapes of these films. R. Murray Schafer, shifting the focus from how we experience the land from one of seeing to one of listening, writes that "hearing is a way of touching at a distance,"[28] but more than that one could surmise that it is also a way of touching across time. *Wuthering Heights* and *Sunset Song* use sound and music as a way of negotiating temporalities. The use of asynchronous sound and the strategy of allowing sound to bleed across scenes provide a way of moving between the past, present, and future. Disembodied voices carried on the wind provide a link to the past, while also foreshadowing events in the future. Indeed, one can surmise that soundscape recording or acoustic ecology functions as a type of hauntology or spectral presence. As Barry Truax writes,

> A striking advantage of the electroacoustic medium, on the other hand, is the layering of what may be called different "levels of remove" where the actual present, the recorded present of the running commentary, the reenacted and remembered past, as well as imagined events past or future, may co-exist with the listener moving fluidly between them.[29]

Sound is manifested through changes in beriatric pressure. It can be experienced on a corporeal level. One is being physically caressed by an embodiment of the past, in the ear drum, on the skin. In the films explored in this chapter, the sound and music form a secret phonography that inscribe traces of hidden narratives on their landscape, affording Derrida's specters a medium with which to express their mysteries.

Schafer proposes the idea of keynote sounds that permeate the acoustic environment. These keynote sounds are not always consciously heard yet, in the words of Schafer, help to "outline the character of men living among them,"[30] creating a sonic identity for a place and its people and suggesting the *possibility* that they can exert an influence on the mood and behavior of those that listen to them. In *Wuthering Heights* the keynote sounds[31] of the wind, the lapwings and acoustamized singing resonate across the landscape. Sounds that are typically hidden from us on a daily basis are foregrounded within the sound mix. We hear the sound of beetles, the buzzing of moth wings, blades of grass blowing in the wind, the traumatic sounds of a haunted landscape rupturing with violence. The sounds of church bells and the disembodied chorus of religious songs permeate the soundscapes of both films, trying to impose order on these unruly carnal landscapes.

Schafer suggest that people are emotionally drawn to specific types of natural soundscapes at particular times of their lives, what bioethicist and composer Bernie Krause refers to as "totem soundtracks."[32] The characters in *Wuthering Heights* and *Sunset Song* are nostalgically drawn to particular sounds haunting the landscape in the face of traumatic events. For Chris in *Sunset song* it is the pastoral sound of an environment on which modern technology has yet to encroach; it is the sound of her lover's voice filling their home. In *Wuthering Heights* it is the acousmatized sound of Cathy singing, the lapwings shrieking, the wind, the grass, the dogs, the visceral crying of the moor. One could also posit that, on a macro level, the filmmakers are drawn to soundscapes with particular sonic qualities. These totem soundtracks, haunted by unseen specters, sculpt the landscape, investing it with narrative and meaning.

Conclusion

To conclude, through the interweaving of shared memories and narratives, geographic locales can be imbued with symbolic significance. Often responding to a traumatic event, upheaval or nostalgia, these landscapes are arguably haunted by what Gilles Deleuze and Felix Guattari refer to as a refrain, a refrain that is carried forward by generations of communities, different in each iteration yet grounded in the original myth. This is manifest in the hyperreal soundtracks of films such as *Sunset Song* and *Wuthering Heights* which serve to fabricate romantic British landscapes, inculcating them with narrative and meaning. Kirmayer suggests that "Traumatic experience is not a story but a cascade of experiences, eruptions, crevasses, a sliding of tectonic plates that undergird the self."[33] Engaging what Vivian Sobchak refers to as the "sense-making capacities"[34] of our bodies, the intense psychogeographical connection between the characters of *Sunset Song* and *Wuthering Heights* and the sonically constructed landscape haunted by spectral presences not bound by spatial-temporal boundaries provide us with a way of constituting meaning for events that reside outside of language, outside of time.

Notes

1 Edward W. Said, "Invention, Memory, and Place," *Critical Inquiry* 26, no.2 (Winter, 2000).
2 Donald W. Meinig, "Symbolic Landscapces: Some Idealisations of American Communities," in *The Interpretation of Ordinary Landscapes: Geographical Essays*, ed. D.W. Meinig (Oxford: Oxford University Press, 1979), 164.
3 The film adaptations referred to here are as follows:
Wuthering Heights (Emily Brontë, 1847) adapted as *Wuthering Heights* (Andrea Arnold, 2011); *Lady Macbeth of the Mtsensk District* (Nikolai Leskov, 1865) adapted as *Lady Macbeth* (William Oldroyd, 2016); *La Salle Terre* (Honoré de Balzac, 1799) adapted as *This Filthy Earth* (Andrew Kötting, 2001); *Sunset Song* (Lewis Grassic Gibbon, 1932) adapted as *Sunset Song* (Terence Davies, 2015).
4 Said. "Invention, Memory, and Place."
5 Svetlana Boym, "Nostalgia and Its Discontents," *The Hedgehog Review* (Summer 2007), 7.
6 Svetlana Boym, *The Future of Nostalgia* (New York: Basic, 2001).
7 Laurence J. Kirmayer, "Landscapes of Memory: Trauma, Narrative and Dissociation," in *Tense Past: Cultural Essays on Memory and Trauma*, ed. P. Antze and M. Lambek (London: Routledge, 1996), 173–98.
8 *Sunset Song* (2015), [Film] Dir.Terence Davies, UK, Luxembourg: Iris Productions, SellOut Picture, and Hurrican Films.
9 Jacques Derrida, *Specters of Marx: The State of the Debt, the Work of Mourning, and the New International* (London: Routledge, 1994).
10 Colin Davis, "État Présent: Hauntology, Spectres and Phantoms," *French Studies* LIX, no.3: 373.
11 Karl Marx, *The Eighteenth Brumaire of Louis Bonaparte* (New York: International Publishers, 1963 [1852]).
12 Davis, "État Présent," 374.

13 Ibid.
14 Derrida, *Specters of Marx*, xvii.
15 Davis, "État Présent," 377.
16 Jacques Derrida, *Ghost Dance*, Ken McMullen. (Channel 4 Films, 1983).
17 Ibid.
18 Ibid.
19 Akira Mizuta Lippit, "EXC05 Derrida, Specters, Self-reflection," in *Ex-Cinema: From a Theory of Experimental Film and Video* (Berkeley and Los Angeles: University of California Press, 2012), 87–105.
20 Mark Fisher, "What Is Hauntology?," *Film Quarterly* 66, no.1 (Fall 2012), 16.
21 Ibid.
22 Jonathan Schell quoted in Andrew J. McKenna, *Violence and Difference: Girard, Derrida, and Deconstruction/*(Urbana and Chicago: University of Illinois Press, 1992), 142.
23 Boym, *The Future of Nostalgia*, 8.
24 Mark Fisher, "Hauntology Now," *K-Punk*, 2006, http://k-punk.abstractdynamics.org/archives/007230.html (accessed July 9, 2019).
25 Jacques Derrida, "Spectographies," in *Echographies of Television: Filmed Interviews*, ed. Jacques Derrida and Bernard Stiegler, trans. Jennifer Bajorek (Malden, MA: Polity Press, 2002), 117.
26 Lippit, "EXC05 Derrida, Specters, Self-reflection," 93.
27 Lucretious, *On the Nature of the Universe*, trans. Ronald Latham (London: Penguin Classics, [C60 BC], 1994), 41.
28 R. Murray Schafer, *The Soundscape: Our Sonic Environment and the Tuning of the World* (Rochester, Vermont: Destiny Book, (1977/1994), 90.
29 Barry Truax, "Soundscape, Acoustic Communication and Environmental Sound Composition," *Contemporary Music Review*, 1996, 57.
30 Schafer, *The Soundscape*, 9.
31 R. Murray Schafer proposes the idea of a keynote sound as one that demarcates the primary tonal character of an acoustic environment, functioning as a constantly present "Background Against Which Other Sounds Are Perceived" (Schafer, *The Soundscape*, 273).
32 Bernie Krause, *The Great Animal Orchestra: Finding the Origins of Music in the World's Wild Places* (London: Profile Books, 2012), 219.
33 Kirmayer, "Landscapes of Memory," 188.
34 Vivian Sobchack, *Carnal Thoughts: Embodiment and Moving Image Culture* (Berkeley: University of California Press, 2004), 8.

10

The Long Trajectory of Death: Justin Kurzel's Screen Adaptation of *Macbeth* (2015)

Danijela Kulezic-Wilson

In his book *The Spectre of Sound*, Kevin Donnelly describes non-diegetic music in film as a seemingly inexplicable, irrational, "demonic force" which possesses both the film and its audience, supplying cinema with a "spectral presence."[1] It is an evocative idea which resonates in various cinematic contexts, not least those that deal with the ghosts of trauma, such as Justin Kurzel's screen adaptation of *Macbeth* (2015). Kurzel's version presents Shakespeare's famous play not as a cautionary tale about the tragedy of ruthless ambition but rather as a story of destructive forces born of unprocessed grief and trauma that bring the protagonists to the point where the object of their longing is not power but death. This reading is brought to life by a directorial style that foregrounds the materiality of the film body through an exceedingly sensuous interplay of image and sound. This particular body, however, is a product of a tortured (fil)mind haunted by apparitions and disturbing thoughts. It vividly conjures Daniel Frampton's idea of film as an organic intelligence ("filmind")[2] that creates events and characters according to its own laws: its ragged wind-beaten landscapes are often tinted with hues of blue, orange, and blood red; dead soldiers roam the fields; the earth trembles in response to killings, while music mourns the characters who commit them. Drawing on Donnelly's description of film music as a force which possesses a film's characters, and Frampton's idea of film as a poetical from of thinking about the world I argue that the score composed by the director's brother, Jed Kurzel, embodies the self-destructive forces that invade the Macbeths like a mental fog, eroding their sanity and integrity, and beckoning them toward death.

Landscapes of Trauma

According to Frampton's filmosophy, filmind should not be understood as a specific, "thinking part" of a film because filmind *is* the film itself, a world with its own intentions and creativities.[3] Filmind does not come from a particular realm or perspective—it is both itself and the character, both objective and subjective.[4] Filmind thinks through images and sounds, creating a world whose events and characters "act out an interaction with the world" so that we can experience it and recognize our own interaction with our world.[5] If this explanation resembles a philosophical take on the relationship between God and the universe, this is not accidental: Frampton himself acknowledges this connection when he writes that our enjoyment of the experience of film may originate in our wish to be part of the world created by an "absolute mind," mirroring the way God might be "busily thinking our world."[6] And like our world, the *Macbeth* filmind also began with the Word (or rather, the Verse), but it does not revolve around it entirely, expressing itself also through nonverbal audiovisual means that are deeply rooted in the images and sounds of the Scottish landscape.[7]

The landscape in *Macbeth* is both the anchor of the film's materiality and the source of the elusive and supernatural. Its sharp, jagged hills often shrouded in fog and the sound of bone-penetrating wind bring a palpable haptic quality to Macbeth's world.[8] The landscape is also the home of witches, and the inspiration for music which represents the destructive impulses haunting the protagonist. Although Jed Kurzel's score is organically connected to the landscape dominating Macbeth's world, as if it's "coming out of the sky, coming out of the mountains,"[9] the music resists any specific references to Scottish folk music. The main musical themes are given to the strings, their melancholic microtonal slides evoking an archaic quality which matches the "earthy, organic, rustic" look of the film.[10] The music is also associated with the witches, which was initially a response to the fact that their appearances are dramatically reduced in the film adaptation compared to the play.[11] The music gives them a "voice" even when they are not on screen, alluding to their "hovering and controlling presence."[12] According to the composer, this was partly the reason he avoided any explicit or implicit references to Scottish folk music as he wanted the score to maintain a certain sense of "foreignness," reflecting the mysterious forces that affect the characters.

In Kurzel's adaptation, however, the idea of the supernatural is not taken at face value and instead becomes the starting point for challenging traditional readings of the characters' motivations. Even though *Macbeth* is generally perceived as a morality tale about voracious ambition, the motivational forces driving the characters in Shakespeare's play are not unambiguous. Are the Macbeths the incarnation of power-hungry evil? Are supernatural forces, represented by the witches, to be blamed for the crimes that ensue? Does Macbeth, faced with the witches' wicked plans, stand a chance of resisting them or is their purpose to expose the protagonist's decaying moral core? Understanding the role the witches play in Macbeth's downfall is an important part of Kurzel's adaptation as well, but he looks for answers about their nature beyond the world of the supernatural. His interrogation of Macbeth's motivation is additionally complicated by the idea that the Macbeths, here played by Michael Fassbender and Marion Cotillard, had a child who died and whose funeral is presented in the opening scene of the film. This detail suggests that Lady Macbeth's involvement in inciting her husband to murder King Duncan is her way of dealing with, or rather suppressing, her grief. Another narrative detail that is given much weight in this screen adaption is Macbeth's participation in a civil war, his subsequent moral and mental decline attributed to his suffering from Post-traumatic Stress Disorder.[13] This theme is, from the beginning, entangled with the notion of the supernatural as evident in the early scene of the battle of Elgin in which Macbeth defeats mercenaries hired to overthrow King Duncan [00:04:25–00:07:51].[14] The scene displays the sensuousness of the film's audiovisual language in its full glory, while the ambiguity of its narrative spaces offers early clues about the corrosive energies that will emerge from its filmind.

The battle unfolds in rhythmically edited images of carnage punctuated by slow-motion shots which sometimes almost register as freeze-frames. These changes in visual rhythm are mirrored in equally contrasting alternations of sound and silence: the sonic onslaughts of shouting, grunting, thumping, and sword-clashing disappear in slow-motion shots, swallowed by a distant humming noise, which is in turn perforated by a burst of sound at the return of the normal projection speed. In addition to their rhythmic function, the slow-motion shots also create an eerie effect, especially in the moments in which both Macbeth and a boy soldier who is about to die notice the figures of three women and a child observing the battle—the witches who are traditionally associated with the supernatural forces pulling the strings of the play. The boy's death will haunt Macbeth throughout the film, suggesting that, after losing his own son, Macbeth's fatherly affections were transferred to the boy. Following the boy's death, the rest of the battle is shown intercut with the scene of the Captain reporting to King Duncan about Macbeth's victory, while the music from the battle scene continues in the background.

The scene's audiovisual flourishes have an underlying narrative function since the fragmented representation of the battle points to the presence of different observers/points of view. While one of those points of view is explicitly attributed to the Captain, it is implied that another one belongs to the witches, the use of slow motion and muted diegetic sound alluding to their otherworldly qualities. This reading is also supported by the framing of the first six shots of the battle: the images of soldiers running and then clashing with the enemy are all presented as medium shots while the slow-motion images intercutting the action are framed as long shots, implying that they are being observed from a distance. The representation of the battle from various perspectives, some of which are deliberately ambiguous, not only undermines the stability of clear-cut narrative spaces but also reflects contradictory aspects of the protagonist. Described by Lady Macbeth at the beginning as being "too full with the milk of human kindness," over the course of the play/film Macbeth undergoes remarkable transformations from a faithful subject of the crown to a regicide, and from a noble warrior to a paranoid murderer, which suggests he has been overcome by mysterious forces.

The music is an integral part of understanding the character's transformation as it is from the beginning linked to the witches in two defining moments marked by the protagonist's intense suffering. The witches are first seen at the end of the prologue which shows the burial of Macbeth's child where they announce they will meet again "When the hurly-burly's done; When the battle's lost, and won."[15] They next appear during the carnage when Macbeth witnesses the death of the boy soldier, another person he deeply cared for. In that sense the witches can be interpreted as the manifestation of trauma, which also leaves space for the possibility that a big part of what we see originates in Macbeth's metadiegetic space.[16] This interpretation is additionally supported by the fact that, both before and after the witches appear in a slow-motion shot during the battle, we see the figures of three mercenaries from the enemy army standing in exactly the same spot, implying that the witches might be a figment of Macbeth's imagination. In any case, the scene of the young soldier's death (which undoubtedly revives Macbeth's memories of losing his own son), followed by the shots of the witches and the shot of Macbeth standing still in the middle of the combat mayhem staring at them, captures the moment of Macbeth receiving the final emotional wound which will fester and contaminate his whole being. This moment is not only separated from the rest of the battle and brought into focus by the use of slow-motion and muted diegetic sound but also by the appearance of music whose mournful character stands in sharp contrast to the violence on screen. The material in both scenes featuring the witches stems from the same origin, ascending and descending microtonal slides (in the prologue they revolve around B-D, D-B flat, B flat-B and in the battle scene it is E-D-B). The important point in this context is that, however we choose to interpret Macbeth's actions and the events around him, both the prologue and the battle scene establish music as being intrinsically connected to his traumatic experiences. Thus, the idea that the music gives voice to the witches, and to the notion that they are "controlling" Macbeth even when they are not on screen,[17] can also be understood as music giving voice to Macbeth's trauma, becoming a sonic embodiment of the destructive forces that will corrupt both him and his wife.

The overlapping of the idea of trauma with the theme of the supernatural presented in hallucinations and apparitions makes a lot of sense when we consider modern theories of trauma which suggest that an individual overwhelmed by extreme events loses the ability to assimilate memories into a linear narrative and is often afflicted by fragmented memories which return in flashbacks and nightmares, blurring the line between past and present.[18] As Judith Herman explains, this kind of fragmentation "whereby trauma tears apart a complex system of self-protection" is often paired with chronic arousal of the autonomic nervous system which can result in irritability and explosively aggressive behavior.[19] Macbeth's actions after his return from war clearly show that he has lost all respect for human life. These actions, however, originate in visions seemingly connected to his PTSD, which apart from the witches also include memories from the battle, as apparent in the scene preceding Duncan's murder ("Is this a dagger which I see before me …"). Following a series of flashbacks of the battle, Macbeth is faced with the apparition of the killed boy soldier who hands him the dagger. In a later scene, after visiting the witches to ask where his bloody deeds will take him, Macbeth finds himself on a field at dawn surrounded by the ghosts of dead soldiers. The final prophecy ("Be bloody, bold, and resolute …"), which in the play is attributed to an apparition of a blood-stained child, is pronounced to him by the same boy who gave him the dagger to kill Duncan.

The music's persistently mournful character, however, seems also to show compassion for the protagonist, as if acknowledging the hopelessness of Macbeth's situation. This is particularly obvious in the scene preceding Duncan's murder. As Macbeth talks about "a dagger of the mind" coming from his "heat-oppressed brain," addressing the boy's apparition, we hear a variation on the theme associated with trauma/witches. The music continues as the boy leads Macbeth toward Duncan's tent. The moment is more somber than dramatic as the music keeps the focus of the scene on Macbeth's tortured mind rather than the drama of the imminent murder. In fact this is the only moment in the film in which the usually constrained microtonal slides develop into a melody, the lamenting sighs in the violin seemingly mourning Macbeth's doomed destiny rather than condemning him.

The Haunting of the Filmind

There's an interesting parallel between the function of the score in Kurzel's *Macbeth* and of the one Jed Kurzel composed for Jennifer Kent's film *The Babadook* (2014). The latter has been described both as a horror and a psychological thriller. On one level it is a version of a "monster" movie where the elusive character of the Babadook terrorizes a single mother Amelia (Essie Davis) and her son Samuel (Noah Wiseman). Toward the end however, it becomes clear that this monster was brought to life by Amelia's denial of grief following the death of her husband. The suppressed sorrow that she refuses to talk about gradually takes over her life, becoming an independent entity, a "creature" that infests her mind, becoming a direct threat to her and her son. What is interesting in this context is that the music in *The Babadook* gives a sonic body to the suppressed grief tormenting the character in the same way the witches visually represent the physical embodiment of Macbeth's disturbed mental state, while the music sonically embodies this disturbance even when the witches are not onscreen.[20]

Both examples resonate with Kevin Donnelly's earlier quoted description of non-diegetic music as a "demonic force" which possesses both the film and its audience. This idea of non-diegetic music being part of the diegetic universe has been given a useful theoretical framework by Ben Winters's idea of intra-diegetic music which also draws its inspiration from Frampton's concept of a filmind as a world of its own emanating "from a uniquely transsubjective *non-place*."[21] Winters argues that music which is usually considered non-diegetic could actually be envisioned "as a kind of energy field imagined by the filmind, which surrounds the space of the film and responds to the presence of characters, or to other elements of the diegesis."[22] Even though intra-diegetic music is "the product of narration" and does not have an obvious source on the screen like diegetic music, Winters contends that it "belongs to the diegesis just as surely as the characters; and, furthermore, it may respond to them, or be shaped by them."[23] This seems like a precise definition of what happens with music in *Macbeth*, *The Babadook*, and many other films; music not only surrounds or responds to the characters, it actually gives body to certain aspects of their psyches. Or, as Donnelly puts it in Freudian terms, "Film music manifests an (not the only) unconscious level of the film, as well as a level of unconscious within the film."[24]

Winters's notion of intra-diegetic is complemented with the term "extra-diegetic music" which is "music or sound whose logic is not dictated by events within the narrative space."[25] In this category, Winters includes "music that accompanies certain montage sequences, or seems to be deliberately distanced from the here-and-now of the narrative space's everyday world."[26] We could argue, though, that this division might need to be reconsidered in films such as *Macbeth* that eschew classical narration in favor of interrogative approaches in which formal reflexivity might overlap with subjective or expressionistic imagery. The battle scene in *Macbeth* supports this argument effectively because, following Winters's logic, the distancing slow-motion shots and their mournful music that take us out of the here-and-now of the narrative space should be interpreted as extra-diegetic. But if we accept them as the point of view of the witches, or even as an insight into Macbeth's inner world as experienced by a character in close proximity to death, its extra-diegetic status would inevitably be revoked.[27] Considering that contemporary narrative methods thrive on ambiguity and contradiction, blurring the lines between subjective and objective, classical and meta-narration, this inevitably affects the status of music within the narrative, which makes the distinction between intra-diegetic and extra-diegetic too ambiguous to be useful.

The unstable lines between narrative spaces in *Macbeth* are certainly an important part of its interrogative strategies. The line between diegetic and meta-diegetic is particularly obscured and is at its most ambiguous in scenes with an asynchronous use of speech. Throughout the film Shakespeare's verses are spoken in non-declamatory fashion and are often delivered in a confessional, intimate manner, but are generally presented through dialogue or as pensive monologues. Following the night of the banquet during which Macbeth is confronted by Banquo's ghost, however, a big part of his subsequent monologue is presented asynchronously, giving the scene a particularly oneiric quality. The sense of subjective point of view is reinforced by the representation of the landscape in the scene that follows, showing Macbeth riding into the wilderness at dawn, his disembodied voice suspended over the fields and mountains which turn increasingly cloudy and misty, infused with a pale green tint [01:04:05–1:06:00].

And yet, even though the film often asks us to question which part, if not the whole film, originates in Macbeth's meta-diegetic space, it is not necessary to know the answer to that question to understand the nature of the bond between the music and the protagonist. Eschewing the conventions of narrative cueing according to which music typically responds to the changes in narrative action and characters' surroundings, the score's persistently mournful character clearly prioritizes the protagonist's subjectivity, reinforcing the idea that his every action, encounter, and decision are colored by or directly arise from one source in the same way the film's individual scenes are tinted with the shades of one dominant color. The accent on sonic textures marked by detuned string instruments and the use of microtonal slides rather than on melody and harmony underlines the elusive and corrupting nature of the forces controlling Macbeth.[28] The subjective aspect of Macbeth's experiences is also depicted through the collaboration of the score and the sound design as in the scene of Duncan's murder. In a film populated with violent deaths, Duncan's murder—which is not shown in the play at all—takes center stage but, like the famous shower scene in *Psycho*, its gore is achieved less through explicit visual means and more through suggestive uses of music, sound, and image editing. The first stab of the knife is not seen but rather *experienced* through a minor sonic explosion which detonates a whole series of sonic events. The distant sounds of earth crumbling effectively capture the outrage of nature in response to the killing of the king, as described in the play, but they also signal the tectonic change in the character who is fully aware of the horridness of his deed and yet is unable to resist it.

A Call from Beyond

The consequences of unprocessed grief and trauma can be extremely corrosive and destructive, as suggested by both *Macbeth* and *The Babadook*. If such powerful emotions are reinforced by excessive exposure to death, as experienced by Macbeth during the civil war, they can lead to annihilation of natural existential instincts, replacing them with survivor's guilt. As Benedict

Nightingale writes, Macbeth "acknowledges his impulses with dread, submits to them half-knowing the consequences, and watches himself destroying himself in a long suicide of the soul."[29]

But even a single death, if it is of a loved one, can be the juncture at which the tacit acceptance of life and its rules crumbles, establishing death as the dominant force. In Kurzel's adaption of *Macbeth* we are also invited to consider Lady Macbeth's actions in light of her grief over losing a child. Kurzel's evident artistic liberty might have been inspired by Lady Macbeth's opening monologue in which her summoning of dark forces includes allusions to breast feeding ("take my milk for gall").[30] His adaption certainly suggests that Lady Macbeth's murderous ambitions arise from her anguish, an attempt to suppress her grief.[31] The immensity of that grief affecting both Macbeths is implied through a number of subtle details such as Macbeth deliberately looking away when Banquo's son comes to greet his father when he and Macbeth return from battle. Another memorable moment is the montage sequence showing Macbeth seemingly agonizing over his wife's plan to murder Duncan and Lady Macbeth preparing a sleeping potion for the king's guards, which is accompanied by a distant sound of children singing, an electronic drone giving it an unsettling character.[32] In the next scene it is revealed that the sound comes from the children's choir performing at a dinner prepared in Duncan's honor that both Macbeths later attend. This is one of many examples of asynchronous sound that blurs the line between objective and subjective, and diegetic and meta-diegetic, connecting different timelines through sonic implication of inner disturbance and undermining any simplistic reading of what we see.

After witnessing the killing of the Macduffs, Lady Macbeth's steely ambition dissipates and her calculating and cold demeanor gives way to repentant mournfulness. This transformation is evident in her final monologue which is traditionally presented in a flurry of guilt-driven ramblings interrupted by a doctor and her servant but which in the film is delivered uninterrupted by Cottiard, with remorseful surrender [01:19:19–01:23:00]. The presence of her dead child in the scene, which is only revealed at the end of her monologue, might not be a very subtle way of pointing to the source of Lady Macbeth's devastation, but it is certainly instrumental in giving a focus and confessional purpose to her speech. Her soliloquy is at first punctuated by long pauses, as if the words are rising from hidden depths, but there is an evident change when she reaches the final stanza—"to bed, to bed; there's knocking at the gate; …"[33] These words suddenly flow with ease, and with tears, and this is the moment when the music emerges. Until that point Lady Macbeth's monologue is spoken without any accompaniment, punctuated by silences, the music only faintly present in the background and hardly audible.[34] The music becomes clearly audible only when the character embraces the idea of death, which can be understood as the filmind's forgiving gesture expressed through empathetic accompaniment. In the context of the music's general relationship with death and trauma over the course of the film, however, it can also be argued that the music is not only here as the filmind's response to the character but has been summoned into existence by her or by death itself. Rising from the concluding verses of Lady Macbeth's monologue, specifically from the words "to bed, to bed," the soothing melody is a lullaby; it lulls her into surrendering to death. The reverberating violin line played *sul ponticello* in a very high register, as fine as a spider web, produces a glassy sheen, hinting at some otherworldly presence which is confirmed when Lady Macbeth walks outside to see the witches waiting for her. The melody in Dorian mode is at first unaccompanied, its harmonic ambiguity providing a subtle layer of tension while sustaining the focus on the monologue. The harmonic resolution that follows is the first notable traditional cadence in the film that signals the emergence of an empathetic turn toward the character. Unlike in the play, the words "to bed, to bed" are repeated once more at the end of the scene when Lady Macbeth starts walking toward the witches, the lines acting not only as a rhythmic coda to her departing monologue but also as a confirmation of the character's surrender to death.

An echo of this approach can also be found in the scene of Macbeth's death. Macbeth's last words spoken during his fight with Macduff are "Damned be him that first cries 'Hold, enough.'"[35] The turning point of this fight though, comes earlier when Macbeth realizes that the prophecy protecting him from anyone "of woman born" does not apply to Macduff because he was "untimely ripped from his mother's womb." Having already subdued Macduff, Macbeth suddenly declares "I'll not fight with thee" and retreats, allowing Macduff to stab him. This moment confirms Macbeth's complete subjugation to the forces that have presented themselves through apparitions and prophecies, but his blind obedience to them also offers him an escape from his tortured existence. After this turning point a simple motif in the violin first emerges unaccompanied then morphs into a cadence in a natural minor, or rather the Aeolian mode, as the composer throughout the film avoids intimations to the traditional minor scale by never using the dominant with the raised seventh and opting for parallel fifths instead of full chords with thirds. Nevertheless, as in the scene of Lady Macbeth's final monologue, which was the first in the film to offer a more stable cadential ending, a sense of resolution only appears when it becomes clear that Macbeth's death is imminent.

The Trajectory of Death

For the Romantics, the certainty of death is what makes life bearable. A brush with death invigorates and can be transformative in a positive way, a reminder that our time is finite and life is precious. The death of a loved one however, can cause lasting devastation. An exposure to multiple deaths, or prolonged contact with death and killing, can also be the death of one's natural existential instincts. Justin Kurzel's interpretation of *Macbeth*, and his authorial interventions that add the motif of parental bereavement and present in gory detail the opening battle that is only mentioned in the play, encourage the reading that the litany of violent deaths populating the play and the film originates in the deaths that precede Macbeth's first crime: the death of a child and the deaths Macbeth both witnesses and inflicts on the battlefield. The Macbeths plot, deceive and murder, trying to run away from their inescapable inner horror. They resist and suppress the force of Thanatos with all their cruel actions but only after finally surrendering to it are they granted some peace.

Antonin Artaud suggests that cinema's power is in its ability to capture "those darker, slow-motion encounters with all that is concealed beneath things, the images [...] of all that swarms in the lower depths of the mind."[36] Kurzel's *Macbeth* is a good example of how these concealed parts of the mind can be represented in images and sounds of the filmind. The landscape in *Macbeth* harbors the memories of all the battles and murders that haunt the protagonist's mind; it houses the witches that give voice to those torments; it also absorbs Macbeth's fears and anguish, dyeing his surrounding in blues, greens, and reds, giving his world a dream-like quality. Jed Kurzel's score, on the other hand, gives a sonic body to the forces of grief, trauma, and the self-destructive urges that haunt the protagonists. The mournful tone of its energy field—to use Winters's expression—is unyielding throughout the film, a persistent reminder that the tragic events unfolding in front of us originate in minds clouded by death. The music also grieves for the characters, even when they are shown at their cruelest, and it humanizes them to the point where their deaths are perceived as genuinely moving. Frampton writes that "Film can think via music as much as noise effects."[37] In Kurzel's *Macbeth* the music seems to know from the very beginning that it is not ambition, or pure evil, or the supernatural that controls the Macbeths, but their longing for death. It knows where the characters are going and it accompanies them and sometimes guides them while they search for their place of rest.

Notes

1. Kevin J. Donnelly, *The Spectre of Sound: Music in Film and Television* (London: BFI, 2005), 20.
2. Daniel Frampton, *Filmosophy* (London and New York: Wallflower Press, 2006).
3. Ibid., 5.
4. Ibid., 87.
5. Ibid., 6.
6. Ibid., 5.
7. In the DVD Extras various members of the production team and actors talk about landscape as an integral part of the story, "connected to the words and to the performance." Jed Kurzel voiced the same sentiment in my Skype interview with him on November 9, 2015 and during the Composition Masterclass he gave at the Cork Film Festival on November 12, 2015.
8. The role of landscape in giving the film a distinctly cinematic quality has been recognized by a number of reviewers, even prompting the criticism that its pictorial splendor might have come "at the cost of violence to the playwright and play, which have been toyed with and sliced up" (Michael D. Friedman, "The Persistence of Fidelity in Reviews of Kurzel's *Macbeth*," *Literature/Film Quarterly* 47, no.4 (Fall 2019), https://lfq.salisbury.edu/_issues/47_4/the_persistence_of_fidelity_in_reviews_of%20kurzels_macbeth.html (accessed December 7, 2022). See also Nigel M. Smith, "Cannes: Justin Kurzel on His Vision for 'Macbeth' and Fascination with Violence," *IndieWire*, May 26, 2015, www.indiewire.com/2015/05/cannes-justin-kurzel-on-his-vision-for-macbeth-and-fascination-with-violence-61536 (accessed December 7, 2022).; Claire Hansen, "Review: Justin Kurzel's *Macbeth*," Shakespeare Reloaded, November 10, 2015, www.shakespearereloaded.edu.au/review-justin-kurzels-macbeth (accessed December 7, 2022).
9. Kurzel, Composition Masterclass.
10. According to Kurzel, the core material for *Macbeth*'s score was created during improvising sessions with five members of the London Contemporary Orchestra

who he previously worked with on *Slow West* (John Maclean, 2015). He says that he treated the musicians as "actors in the scene," asking them to experiment and improvise, searching for sounds and textures that felt part of the landscape while also bringing a sense of "foreignness." (Skype interview with the author; Composition Masterclass).

11 Kurzel, Skype interview with the author.

12 Kurzel, Composition Masterclass.

13 Henry Barnes, "Michael Fassbender: Macbeth suffered from PTSD," *The Guardian*, May 23, 2015, https://www.theguardian.com/film/2015/may/23/michael-fassbender-macbeth-suffered-from-ptsd.

14 This is the timing of the scene in the DVD release of *Macbeth*, Studiocanal, PAL, 2016. The duration of film and timings of the quoted scene may differ between PAL and NTSC formats and also depending on which program is used to play the DVD.

15 William Shakespeare, *Macbeth*, ed. Cedric Watts, Act 1, Scene 1 (London: Wordsworth Classics, 2005), 31.

16 This reading is also supported by something Jed Kurzel mentioned during our Skype conversation. Responding to my comment that in his brother's adaptation the speech often seems decentralized he said that he always thought about the story as being Macbeth's dream. Referring to the film's use of asynchronous speech he said that it reminded him of how people sometimes talk in a dream, "almost like they are talking dead—they are talking but they are not moving their mouth."

17 Kurzel, Composition Masterclass.

18 See for instance Judith Herman, *Trauma and Recovery: The Aftermath of Violence—From Domestic Abuse to Political Terror* (New York: Basic Books, 1992); Maria Cizmic, *Performing Pain: Memory and Trauma in Eastern Europe* (Oxford: Oxford University Press, 2011).

19 Herman, *Trauma and Recovery*, 35–6.

20 Although the "horror" aspect of the film rests on the idea that the malevolent force tormenting Amelia's son comes from the children's book with a character named Babadook, the first indication that the source is situated somewhere else, something that Samuel seems to be aware of all along, comes in the scene showing Amelia after an argument she had with Samuel because he scribbled all over a photo of his late father. Their confrontation ends with her son shouting "do you wanna die?" accompanied by a single ominous sound which continues into the next scene showing Amelia sitting at the table, visibly upset. As we see Amelia massaging her head, the gradual buildup in the music achieved through layering of rhythmic motifs over the original drone, intercut with images of a lightbulb flickering and Samuel staring fearfully into the darkness, breathing heavily, signals the appearance of the "monster." The focus of the scene, however, is on Amelia's head which is repeatedly shown from different angles, the camera coming closer

to her in each shot, as if trying to localize the source of the disturbance. Although the meaning of this framing and of Samuel's shouts "don't let it in, don't let in!" can be fully comprehended only at the end of the film, the fact that the source of the musical build-up seemingly originates within Amelia's head, or rather her state of mind, mirrors the function of the music in *Macbeth*.

21 Frampton, *Filmosophy*, 86.
22 Ben Winters, "The Non-diegetic Fallacy: Film, Music and Narrative Space," *Music and Letters* 91, no.2 (May 2010): 242.
23 Ibid., 234.
24 Donnelly, *The Spectre of Sound*, 21.
25 Winters, "The Non-diegetic Fallacy," 237.
26 Ibid.
27 Winters's argument is mostly based on his analysis of classical narrative and scoring but even in this context his explanation of the difference between intra-diegetic and extra-diegetic might seem problematic. In order to illustrate the notion of extra-diegetic Winters gives the example of Sgt. Elias's death in *Platoon* (Oliver Stone, 1986) where Samuel Barber's *Adagio* "seems distanced from the narrative action" (237). While I agree with his proposal that the music in this scene may be perceived "as an expression of the filmind's own emotional reaction" (237), I am not convinced that this reaction should be considered separate from the narrative space because if intra-diegetic music is "a kind of energy field imagined by the filmind" why would the filmind's emotional reaction to a certain action be considered separate from it? While the music of *Macbeth* seems to emanate from some place bigger than any individual character, it is nevertheless part of that world.
28 Kurzel, Composition Masterclass.
29 Benedict Nightingale, "Something evil this way comes…," *The Times*, January 21, 2005, 20.
30 Shakespeare, *Macbeth*, Act 1, Scene 5, 60.
31 Justin Kurzel mentioned in several interviews that both his first film, *Snowtown* (2011), and the adaptation of *Macbeth* were influenced by the process of grieving his father's death: "I've had a difficult time decoding my grief, working out what to do with it. That's what I saw in *Macbeth*. They [the Macbeths] use ambition to replace grief." Danny Leigh, "*Macbeth* Director Justin Kurzel: 'You're Getting Close to Evil,'" *The Guardian*, September 24, 2015, http://www.theguardian.com/film/2015/sep/24/macbeth-director-justin-kurzel-australian-film-maker-snowtown (accessed December 7, 2022). See also Smith, "Cannes".
32 For an in-depth reading of *Macbeth* that focuses on the role of children in Shakespeare's tragedy and their relationship to trauma and vulnerability, see Hanh Bui, "Effigies of Childhood in Kurzel's *Macbeth*," *Literature/Film Quarterly* 48, no.1

(Winter 2020), https://lfq.salisbury.edu/_issues/48_1/effigies_of_childhood_in%20kurzels_macbeth.html (accessed December 7, 2022).
33. Shakespeare, *Macbeth*, Act 4, Scene 1, 92.
34. Jed Kurzel says that he liked the idea that the faint and unsteady presence of music in the first part of the scene sounded like it might be playing outside and "you could almost not make out whether you were hearing it or not, whether it was actually there or not" (Skype interview with the author).
35. Shakespeare, *Macbeth*, Act 5, Scene 8, 102.
36. Antonin Artaud quoted in Frampton, *Filmosophy*, 66–7.
37. Frampton, *Filmosophy*, 121.

11

Haunted Folk: Specters of the Analogue in *Annihilation* (2018)

John McGrath

Alex Garland's *Annihilation* (2018) is a film haunted both by its characters' traumatic pasts and the specters and traces of analogue technology. On the surface, an apocalyptic tale of alien invasion and obliteration, at a deeper level *Annihilation* offers a sustained metaphorical allegory exploring what the director sees as the innate human drive for self-destruction. It invokes themes of mental health, infidelity, terminal illness, marriage breakup, and alienation, as the nation (the United States and eventually the world) becomes itself an alien nation. A "weird" cosmic-folk soundtrack evokes the hauntings of the characters via the mirror or "the shimmer" (the name of the alien zone) of the analogue.[1] Indeed, the very idea of analogue technology becomes an analogue too for the dominant theme of the film: the self-destructive impulse of humanity. Replaying and reconfiguring the mistakes and painful memories of the protagonists, this refraction acts like a plate reverb unit, creating fragmented echoes which abound the filmspace. *Annihilation* is not explicitly *about* "technostalgia" or the fetishization brought about by binary polarization of tech/analogue tropes per se; instead, the film and its soundtrack employ these mediated semiotic codes as a functional narrative device.[2] In this chapter, I explore how music affords these themes a holistic resonance.

Genre and *Annihilation*

Annihilation is arguably a hybrid genre film, demonstrating characteristics of both dystopian science-fiction and folk horror film traditions. Dystopian sci-fi is well known for its philosophical metaphorical affordances. *Annihilation* charts, in great depth and with precise attention to detail, the inevitable ruin/transfiguration of Earth by extra-terrestrial biological forces. Both human and alien protagonists must either face up to their *doppelgängers* (doubles)—themselves in mirror form, their dark Jungian shadows, their trauma—or each face annihilation. The alien entity learns of self-destruction from its interaction with the human, ultimately allowing itself to be destroyed so that it may live on in a changed form. As this chapter will discuss, the film is a study in self-reflection so that positive change and transformation can be affected, and trauma overcome/survived.

Annihilation might also however, be viewed as a folk horror film.[3] One of key characteristics of folk horror is its engagement with notions of trauma, of characters haunted by their pasts, and the dangerous disinterring of dark memories. Extending Derrida's notion of hauntological traces, Simon Reynolds writes that "[h]auntology is all about memory's power (to linger, pop up unbidden, prey on your mind) and memory's fragility (destined to become distorted, to fade, then finally disappear," and folk horror highlights the dangers of disinterring the past and of traumatic memories that "prey on the mind."[4] Folk horror works as a useful template to foreground a skepticism of nostalgic, Romanticized or idealized pasts.[5] *Quatermass and the Pit* (BBC, 1958–9) is an obvious pre-cursor to *Annihilation* here, as it also involves the release of sinister forces from a crash-landed alien spaceship. Garland's film *Men* (2022), more overtly references the genre characteristics of folk horror, with a sinister terrestrial/rural setting reminiscent of those found in British folk horror films such as Piers Haggard's *Blood on Satan's Claw* (1971) and Ben Wheatley's *A Field in England* (2013).

Music, The Analogue, and The "Mistake"

> Kane: I was just looking at the moon. It's always so weird seeing it like that in the daylight.
> Lena: Like God made a mistake. Left the hall lights on.
> Kane: God doesn't make mistakes. That's … somewhat key to the whole "being a god" thing.
> Lena: Pretty sure he does.
> Kane: You know he's listening right now, don't you?
> Lena: You take a cell, circumvent the Hayflick limit, you can prevent senescence.
> Kane: I was about to make the exact same point.

Lena: It means the cell doesn't grow old; it becomes immortal. Keeps dividing, doesn't die. They say aging is a natural process, but it's actually a fault in our genes.

Kane: I get really turned on when you patronize me. It's really hot.

Lena: Without it, I could keep looking like this forever.

Kane: Oh, okay. Well, that could constitute a mistake (*Annihilation*, 2018).

The score for *Annihilation* is composed by Garland's long-term collaborators Ben Salisbury and Geoff Barrow, both of whom have also worked on other Garland films and television shows including *Ex Machina* (2014), *Devs* (2020), and *Men* (2022). The duo match Garland's thoughtfulness and depth of detail in their unique approach to the sound world of the film. The film is split into two parts sonically: the analogue and the digital. This binary opposition highlights the otherness of the invasion and the mystery of its nature, playing on standard tropes of analogue authenticity, "technostalgia," and what Peter Narváez calls the "myth of acousticity."[6] The analogue instrumentation works as a metaphor for humanity and the flaws therein, particularly this "drive" for self-destruction. In an interview with Salisbury that I conducted as part of another collaborative project the composer states:

> After we read *Annihilation*'s script and gathered a sense of the film's surroundings and environment, we decided we'd have no electronics except for that last third act where we're in the world of alien insanity, and then all bets would be off. Nothing in the first two acts is produced by a synth, it's all bowed waterphones and bells, as well as acoustic guitar. And this somehow matches *Annihilation*'s color palette.[7]

Therefore, with no electronics for two-thirds of the film its composers are employing tropes of analogue semiotics, engaging with the mediated signs of technological polarization that have pervaded certain genres in popular music history such as lo-fi and vaporwave. The false binary opposition of digital/analogue provides both an artistic palette and a narrative device for the composers to further the film's themes. I suggest that this usage evokes the mediated "humanity" often associated with material gear, tangible platforms, and haunted instruments—the "warmth" of vinyl, the "crackle and hiss," the "natural overdrive" (amongst other hauntological signifiers).

In *Annihilation*, this trope functions to convey the overarching theme of human-error—"the mistake"—due to the fact the fetishization of analogue technology often relies and focuses on the "mistakes," the "decay" of its medium, and the corporeal/spatial "liveness" of its creation.[8] Of course, the fact that Digital

Audio Workstations (DAWs) and effects processing takes place throughout the full score, despite any supposed dialectical construction, is emblematic of the fact that such dichotomies are always illusory. Consequently, in this chapter I focus first on the employment of Americana/Brit-folk guitar and second, on the use of the song "Helplessly Hoping" (1969) by Crosby, Stills and Nash on the film soundtrack. Analyzing these two specific instances of analogue instrumentation in the film multimodally allows for an exploration of instances of "weirdness" brought about through cognitive dissonance, in addition to moments of synchronicity between the visuals and music.

"All along the watchtower"

Kevin Donnelly has suggested that "[l]ike a spectre, film music is disembodied and denies the logic of the rest of the diegetic film world."[9] The nondiegetic music at the opening of *Annihilation* is certainly ghostly in the sense of it being acousmatic, "a specter" in the film world. This sonic haunting goes further, however. We are not accustomed to folk music in space, making this choice stranger than any orchestra, string quartet waltz, or Ligeti microtonal washes. The acoustic guitar instrumental that plays over the opening shots of *Annihilation* works in what Mark Fisher would call a "weird" fashion. For Fisher, the weird involves an intrusion of the "outside" into the "inside."[10] This invasion invokes "a sensation of *wrongness*: a weird entity or object is so strange that it makes us feel that it should not exist, or at least it should not exist here."[11] Further to this, Fisher writes that "the weird is constituted by a presence—the presence of *that which does not belong*. In some cases of the weird ... the weird is marked by an exorbitant presence, a teeming which exceeds our capacity to represent it."[12] Beginning with the "Skydance logo" in the opening credits of the film, we have spectral analogue music comprising of bowed waterphones and bells accompanying steel-string guitar fingerpicking as an extra-terrestrial crashes into the Earth. An alternate-bass picking pattern in DADGAD (6–1) modal tuning is reminiscent of the fingerstyle of Brit-folk pioneers Davy Graham and Bert Jansch combined with a touch of John Fahey's dissonant "American-primitive" style. The theme is in D Dorian (natural minor with #6) at a loose 70 bpm. The soundtrack album lists the piece as "Watchtower," perhaps in reference to the lighthouse in the film. The doubling of the alternating bass notes (thumbed octave Ds) is pertinent to the theme of duplication within the film and indeed a variation of this theme often appears in the film alongside images of cells duplicating and repeating. This juxtaposition of the folk and the alien within the soundtrack is thus stated from the very opening of the film and pervades throughout. The binary opposition of human (folk and analogue aesthetics)/alien (tech) and the *weirding* effect of such is furthered of course by the deliberate play with tropes of corporeal authenticity present in the film, as discussed by Salisbury above. Weaving an avantfolk texture, in which supposed folk-aesthetic tropes are themselves weirded, this opening guitar theme fades out with the line: "this is a cell."[13]

The intrusion of "the weird" is an apt choice to depict an atmosphere of invasion and alienation. Salisbury reflects on the choice as follows:

> But in a film, there's often a weird symbiosis between moving picture and music. *Annihilation*'s opening has a shot with a comet coming through space, and after four or five attempts, we'd settled on one serviceable fragment with waterphone that had a strange atmosphere—it was soundscapey. At 2/3 into the film, we'd written this folksy, backwoods, simple Americana guitar theme to accompany the team walking through the woods in the American South, as well as Lena contemplating things. Alex phoned one day and said, "Oh, I've put the guitar theme over the front, and I think it works." Geoff and I put our heads in our hands, and he said, "No look, just try it." and he sent us a QuickTime immediately. We both went, "Oh my God, yeah he's right." Alex's take on it was, in the film's opening, we've got a captive audience and some latitude to do this. Composers and a filmmaker can tease out an audiovisual relationship that just works: it can be upsetting or satisfying, it can lead you down the garden path, it can be just purely beautiful.[14]

Here, then, is an occasion in which the directorial instinct led to a surprising chance procedure, an unexpected experiment that produced new and unusual results. Part of this apropos weirdness is achieved via the supposed tropes of analogue authenticity at play. We have a haunted space landscape, half-remembered narratives, and specters of humanity. The Eisensteinian cognitive dissonance achieved by montage is part of this weird effect/affect.[15] The dissonance of acoustic instruments over sci-fi backdrops is at times reminiscent of those of *2001: A Space Odyssey* (Kubrick, 1968), but this dance of death is a blues, not a space waltz à la Strauss, and it is perhaps even more ominous. Fahey, who himself explored montage and *musique concrète* on early records like *Requia* (1967, Vanguard) is also brought to mind, as is Fahey's "Dance of Death" (a famous instrumental composition of his from 1965); indeed, the elegiac

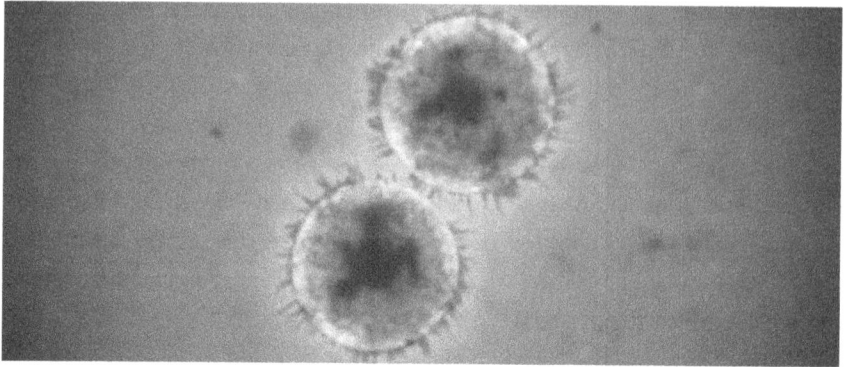

Figure 11.1 Cells Divide in "Official Trailer," *Annihilation* (Paramount Pictures, 2018).[16]

theme of much blues suits the themes of *Annihilation*. Self-destruction and the blues are long partners of course; it is a well-established trope in blues lyrics and mythology stretching from Robert Johnson to Fahey. In such a way, this prelude of folk/blues at the opening of *Annihilation* acts as a non-diegetic overture, an adumbration of the personal devastation and ensuing global decimation.

As the alien crashes into the lighthouse the highly compressed instrumental folk guitar part introduces the title of the film and our first view of the cells dividing audiovisual trope (Figure 11.1). In fact, "Cells Divide" is the soundtrack title given to a more minimal variation of this guitar theme used later in the film (Figure 11.2). This track is at a slightly faster tempo (a loose 77 bpm) but retains a drop-tuned low D alternate fingerpicking folk/blues style in D Dorian with a prominent i-III movement later on in the track.

The theme's analogue (acoustic steel-string guitar) shadows resound throughout the duplication scenes of the film. I will now chart some of the key reiterations of this guitar theme throughout the film. At 32:10, the "Cells Divide" variation is heard as the crew encounter a duplicated deer in a wood within the shimmer. At 50 minutes in, the acoustic "Cells Divide" theme reappears again in Lena's recollection of her husband, Kane, leaving for the trip, a pivotal moment in which the couple separate for the last time, dividing forever *as humans*. We cut straight to a microscopic image of a cell dividing, which further reifies the metaphor. At 1:07:39 we hear the "Cells Divide" guitar theme again, at the start of the "mistake" flashback, to be further discussed later. Thus, when we see images of disjunction and division, be they human or beast, we hear echoes of the folk guitar theme.[17]

Figure 11.2 Author transcription of "Cells Divide" alternate fingerpicking pattern excerpt.

"Two alone, three together, for [four] each other"

At the start of the film Lena is teaching a biology class about a cancer cell:

> Lena: Like all cells, it is born from an existing cell. And by extension all cells were ultimately born from one cell. A single organism alone on planet Earth, perhaps alone in the Universe. About four billion years ago, one became two, two became four.

In musematic fashion, a musical motivic cell too can prove fundamental, as does the fingerpicking pattern in figure two above. The second use of music in *Annihilation* that I analyze is the prominent employment of Crosby, Stills and Nash's "Helplessly Hoping." The song is pivotal to the film and many scenes are edited to it, with key moments of audiovisual synchronicity. We first encounter the song as Lena paints the marital bedroom (5:16), "helplessly hoping" for her husband Kane's return (her Harlequin). The lyrics are matched with framed photographs of Kane in the army and with her. At the lyric "hello" Kane enters. He climbs the staircase in synchronization with the lyric "[s]tand by the stairway" (Figure 11.3). Lena cries as they embrace to the lyric "chocking on hello." In the hug they are "one person" and no longer "two alone." However, this becomes further complicated as the film progresses. On the line "three together" the song is cut with a resonant delay, implying a third figure has entered the relationship, but is it Dan (Lena's colleague with whom she is having the affair) or Kane's *doppelgänger*?

Later, Lena has a flashback to the house and the song returns, concretizing its position as a leitmotif for this location. This time appears diegetically in her living room, emanating from an unseen sound system (1:22:25–49). The song affords a deep audiovisual exploration of marriage breakdown via self-destructive impulses. Kane takes on this dangerous job after discovering Lena's infidelity and in this sense, the film is a fitting allegory for this journey

Figure 11.3 Kane returns from *Annihiliation* (Paramount Pictures, 2018).

and the transfiguration of their relationship.[18] The key scene in which Lena tells Dan that their affair was a "mistake" (in a memory at 107 minutes) points to this:

> Lena: This was a mistake.
> Dan: You think he knows?
> Lena: Yes.

A key moment occurs when the psychologist Dr. Ventress states: "[a]lmost all of us self-destruct … We destabilise the good job. The happy marriage." The house is mirrored from Lena's memory when she enters the shimmer, the staircase of the facility inside the zone a direct reference to their own home. Lena goes on a metaphorical journey, fueled by guilt, mimicking the devastating path that Kane suffered because of her infidelity. All the characters who enter the shimmer are damaged (Figure 11.4). We later learn that Dr. Ventress (Jennifer Jason Leigh), the psychologist, has cancer and resigns herself never to return from the shimmer. She chooses the candidates for the mission into the zone, focusing on those with specific personal problems for the team. Anya Thorensen (Gina Rodriquez) is the paranoid paramedic with a history of addiction, Josie Radek (Tessa Thompson) is the physicist with a history of mental health issues who succumbs willingly to a transformative death, Cass Sheppard (Tuva Novotny) is the grieving anthropologist who has never been whole since the tragic death of her daughter from leukemia.

Garland explained that his interest in VanderMeer's novel, on which the movie is loosely based, was "born out of a funny kind of preoccupation … that everybody is self-destructive."[19] The shimmer soaks up the pathologies and plagues of the characters it envelops; guilt, grief, depression, and paranoia are all transformed in the prism, refracting what it means to be human, just as it does

Figure 11.4 The crew approach "The Shimmer" from *Annihilation* (Paramount Pictures, 2018).

with light and DNA. The central characters are haunted by their mistakes and trauma, while the sound palette is weirded by anachronistic and out-of-place folk music. The "folk," furthermore, are not always human but *become* alien. The dreamlike landscape is populated with the conjured diaphanous specters of its visitors' thoughts, a haunted alien landscape. In the prismatic world of the shimmer, everything fragments and duplicates. Dr. Ventress provides a pivotal commentary later in the film when she quotes Samuel Beckett's *Molloy* (1951):

> Dr. Ventress: It's the last phase. Vanished into havoc. Unfathomable mind, now beacon, now sea.
> Lena: Dr. Ventress?
> Dr. Ventress: Lena? We spoke, what was it we said? That I needed to know what was inside the lighthouse. That moment's passed. It's inside me now.
> Lena: What's inside you?
> Dr. Ventress: It's not like us, it's unlike us. … I don't know what it wants, or if it wants, but it will grow until it encompasses everything. Our bodies and our minds will be fragmented into their smallest parts, until not one part remains … Annihilation. (132:12–55)

The alien force soaks up the subconscious of its visitors. The mirroring of Lena's house within the shimmer is a key example of such pathologies being mapped out, the psychogeography of the place is constructed in collaboration with the visitor, as the water in her glass refracts earlier on. The line from "Helplessly Hoping" recounting a figure "heartlessly helping himself to her bad dreams" is an apt description of how the shimmer operates, feeding itself on the negative thoughts of its inhabitants.

The film itself is haunted by the book, almost remembered. For Garland, the film was "a memory of the book" rather than a scene-by-scene adaptation: "reading it was a dreamlike experience."[20] This approach of course adds to the peculiar ambiguity of the film. Fisher's concept of "the weird" further resonates with the idea of "strange simultaneity," an atemporality and a disinterring of the past, just as the characters' pasts are dug up and replicated in transfigured forms. Therefore, by the end of the film we have a copy of both Lena and Kane, "for [four] each other." In marriage "they are one person," with infidelity "they are two alone," with Kane's *doppelgänger*'s arrival "they are three" (Kane directs his copy to "Find Lena" and the alien responds "I will") and by the end "they are for [four] each other" (both protagonists have been copied/transformed). The progressive entries of vocal harmonies in the song reinforce this effect. Something new arises after the annihilation of the relationship, both Kane's *doppelgänger* and a changed Lena, most notably indicated by the alien tint to her eyes in the final scene's embrace, as they embark on a fresh journey.

In contrast to the analogue "human" aspects of the film's themes, our first encounter with the alien in the lighthouse is marked by foregrounded electronic synthesis. We hear a step sequencer when Lena approaches the crater hole within the building (adorned with Mandelbrot fractals) leading down to an alien Giger-esque, subterranean landscape. As the flower spews and sucks a drop of Lena's blood we hear static sound design mimic the action. An extant piece, "The Mark" by Moderat (*II*, 2013) is employed as a leitmotif for the alien *doppelgänger* itself. This scene is notably edited to the distinctive pre-existing electronic (tech/alien) track.

Conclusion

In the shimmer we have a haunted alien landscape, a place with agency. Like Tarkovsky's "zone" in *Stalker* (1979), the site acts as a key narrative device that enables the characters to enter an altered reality, one that acts upon them and affords metamorphosis. The central couple are tormented by their broken relationship while the sound palette is weirded by out-of-place music, a haunted folk and folk that haunts. The film itself is a transmogrified, ghostly adaptation of "a memory of the book." The analogue/digital dichotomy or false binary is employed and evoked in montage, building on accustomed clichés of acoustic authenticity, what Narváez calls the "myth of acousticity."[21] The shimmer creates a spectral land where echoes abound. This cycle of destruction and renewal, this dance of death, is given great depth by juxtaposing the somatic and organic with the human and the folk/blues mythos. Lena's bruise (first seen developing at 38 minutes) later becomes an ouroboros, the same symbol found on the dead soldier within the facility, and also on Anya's arm earlier in the film. This image of a snake eating itself, the idea of creation emerging from obliteration, reverberates throughout.

The Deleuzian idea of positive transformation through repetition is Lena's revelation. Deleuze reminds us that, at the moment of reception, repetition is protean.[22] For Adrian Parr "Deleuze encourages us to repeat because he sees in it the possibility of reinvention, that is to say, repetition dissolves identities as it changes them, giving rise to something unrecognizable and productive."[23] Derrida too recognized how the present is inevitably haunted by traces of both the past and the future.[24] *Annihilation* is therefore, in the end, less Freudian trauma and more creative *becoming*: from silence comes sound, from destruction rebirth. Lena absorbed something of Anya in the ouroboros, the alien realized it had to destroy itself to grow too, to become something new or perish. In the closing scenes Lena recognizes this regenesis. When her interrogator (Lomax) says that the alien was "mutating our environment, it was destroying everything" Lena responds: "It wasn't destroying, it was changing everything, it was making something new." Lena at last sees the terrible beauty in repetition and transfiguration:

> Lena: The mutations were subtle at first; more extreme as we got closer to the lighthouse. Corruptions of form. Duplicates of form.
> Lomax: Duplicates?
> Lena: [She looks at the tattoo of ouroboros on her arm] Echoes.
> Lomax: Is it possible these were hallucinations?
> Lena: I wondered that myself ... but they were shared among all of us. It was dreamlike.
> Lomax: Nightmarish?
> Lena: Not always. Sometimes it was beautiful. (37:50–38:17)

Notes

1   Mark Fisher, *The Weird and the Eerie* (London: Repeater, 2016).
2   Trevor Pinch and David Reinecke, "Technostalgia: How Old Gear Lives On in New Music," in *Sound Souvenirs: Audio Technology, Memory and Cultural Practices*, ed.

Karen Bijsterveld and Jose van Dijck (Amsterdam: Amsterdam University Press, 2009), 152–66; Kim Bjorn and Scott Harper, *Pedal Crush: Stompbox Effects for Creative Music Making* (Copenhagen: Bjooks, 2020), 362.

3   The most blatant bit of horror in *Annihilation* is of course the mutant "bear" scene. Incidentally, this creature was designed by the same effects guru who made Paddington in the recent films (this bear is playfully named Homerton, after another London tube station—one more "rough-around-the-edges") https://ew.com/movies/2018/02/25/annihilation-screaming-bear/ (accessed September 15, 2022).

4   Simon Reynolds, *Retromania: Pop Culture's Addiction to Its Own Past*, 1st American ed. (London and New York: Faber & Faber, 2011), 335.

5   John McGrath "'Something Seems Wrong, Should That Be Happening': Avantfolk Guitar and Glitch Aesthetics, A Practice-Based Perspective," in *21st Century Guitar: Evolutions and Augmentations*, ed. Richard Perks and John McGrath (New York: Bloomsbury Academic, 2023), 185–204.

6   Pinch and Reinecke, "Technostalgia," 2009; P. Narváez, "Unplugged: Blues Guitarists and the Myth of Acousticity," in *Guitar Cultures*, ed. Andy Bennett and Kevin Dawe (Oxford: Berg, 2001), 27–44.

7   Holly Rogers, John McGrath, Carol Vernallis, and Dale Chapman, "Composer Ben Salisbury Discusses Scoring Science for Alex Garland," in *Cybermedia: Explorations in Science, Sound, and Vision*, ed. Holly Rogers et al. (New York: Bloomsbury Academic, 2021), 141.

8   Reynolds, *Retromania*, 335; Paul Sanden, *Liveness in Modern Music: Musicians, Technology, and the Perception of Performance* (London & New York: Routledge, 2013).

9   K. J. Donnelly, *The Spectre of Sound: Music in Film and Television* (London: BFI, 2005), 9.

10 Fisher, *The Weird and the Eerie*, 10.

11 Ibid., 15.

12 Ibid., 61.

13 What I term *avantfolk* is discussed in John McGrath, *21st Century Guitar*.

14 Rogers et al., "Composer Ben Salisbury Discusses Scoring Science for Alex Garland," 141.

15 Sergei Eisenstein, *Film Form: Essays in Film Form*, ed. and trans. Jay Leyda (1949; New York: Harcourt Brace & Company, 1977).

16 "Official Trailer: Annihilation (2018)," https://youtu.be/89OP78l9oF0 (accessed September 15, 2022).

17 *Doppelgängers* and doubles have of course been long-standing tropes in science fiction from *Back to the Future* (dir. Robert Zemeckis, 1985) to *Primer* (dir. Shane Caruth, 2004) and *Looper* (dir. Rian Johnson, 2012) via *Terminator 2* (dir. James Cameron, 1991).

18 Oscar Isaac (Kane) speaks of the destruction of the marriage as a theme in a video interview, https://youtu.be/-eVGKyrl1Po (accessed September 15, 2022).
19 Garland qtd. in an interview for *The Verge*, https://www.theverge.com/2018/2/21/17029500/annihilation-ex-machina-director-alex-garland-sci-fi (accessed September 15, 2022).
20 Garland in video interview https://youtu.be/nYhT5Ey42gg (accessed September 15, 2022).
21 Narváez, "Unplugged."
22 G. Deleuze, *Différence Et Répétition, Bibliothèque De Philosophie Contemporaine* (Paris: Presses Universitaires de France, 1968).
23 Adrian Parr, *The Deleuze Dictionary* (New York: Columbia University Press, 2005), 225.
24 Jacques Derrida, *Specters of Marx: The State of the Debt, the Work of Mourning, and the New International* (1994; London: Routledge, 2006).

12

Sonic Novelty and Conceptual Obscurity: Music, Landscape and Enigma in *Picnic at Hanging Rock* (Peter Weir, 1975)

Jady Jiang

Picnic at Hanging Rock (*Picnic*) is an Australian mystery film narrating a story about a group of Victorian schoolgirls going for a picnic at a remote mountain outcrop called Hanging Rock. However, mysteriously three students and one teacher disappear during the outing. *Picnic* is regarded as a significant example of Australian New Wave cinema,[1] a movement that emphasized Australia's enigmatic geographical landscapes. Indeed, these concerns and an engagement, directly or indirectly, with the country's colonial past are evident in Australian films more generally. These preoccupations can be located in the prevalent theme of "lost children," which normally incarnates a story of white people (often children) lost in the bush,[2] indicating the dysphoria of uncertain cultural belonging for white people who live in Australia. Peter Pierce notes that the lost child in Australian cinema "represents the anxieties of European settlers … The Child stands for the apprehension of adults."[3] *Picnic* is a discernible example of this, wherein the national and cultural "lost" culminates in the personal disappearance of the girls. The unique history of Australia and its colonization by Britain in the late 1700s endows some Australian films, such as *Wake in Fright* (Ted Kotcheff, 1971) and *The Last Wave* (Peter Weir, 1977), with an ambiguous sense of human identity and mysterious topography.

When director Peter Weir was asked about the ending of the film, "What happened to these girls? … Do you think they fell down a hole? … They were abducted by aliens?"; he answered, "Any of these above."[4] He does not provide an explicit answer at the end of the film, instead leaving viewers fearful about the mysterious phenomena of nature. As Jonathan Rayner notes the unfathomable landscapes of the film are rendered through "time-lapse, slow-motion, unexplained voice-over and omnipotent revelation."[5] Although it is a consensus among many scholars that *Picnic* is essentially a scenario of ambiguity,[6] some scholars offer discursive explanations here, with the landscape of Australia—and particularly its natural environment—central to the discussion. For example, Suzie Gibson states that, "[t]he ancient landscape of Hanging Rock itself is deeply evocative of primeval fears of being overwhelmed by something beyond the human."[7] Accordingly, there may be unknown inscrutable power beyond human beings' cognition and control in nature, such as the *genius loci*, the spirit that protects a particular place. The potential multiple explanations of *Picnic* emphasize its ambiguous and inexplicable narrative schemas.

What, however, is the role of music in delineating the haunted landscape and inexplicable riddles in *Picnic*? Does the film music complete or intensify the enigmatic sense of this film when other film elements do not offer sufficient information? If so, in what ways? As Bruce Johnson and Gaye Poole state, "Many of Weir's meanings are generated through sounds: musical, synthetic, natural, domestic, industrial."[8] However, the sonic aspect of Australian cinema has arguably been paid scant attention.[9] This chapter will explore the duality in the soundtrack of *Picnic* between classical music and exotic instruments, especially the panpipe. This is a notable sonic opposition that defines the structure of the film; setting up a conflict between civilization and the landscape, colonial society, and the untameable outback.[10] It is clear that the film juxtaposes images of middle-class colonial culture with those of imposing natural landscapes, but this process is perhaps even more clear in the film's dual usage of the different forms of music. Whilst the classical music refuses to "merge" with the dusty Australia outback, the panpipe music appears more enmeshed with the landscape, exhibiting a contrasting incongruity with classical music.

Classical Music: Conventional Sonic Background and Exotic Audiovisual Oscillation

As Adorno and Eisler note, classical music in cinema is normally regarded as "worn-out musical pieces,"[11] which they call musical clichés that are likely to be outmoded, uneventful, and over-familiar, yet remain powerful. The use of classical music stands is conspicuous in *Picnic*, including Bach's Prelude No.1 in C major from *The Well-Tempered Clavier* (1722), Beethoven's Piano Concerto,

No. 5 (the "Emperor," 1809), and Mozart's *Eine Kleine Nachtmusik: Romance* (1787). I would argue that classical music in *Picnic* functions as:

(1) Period-setting. Classical music corresponds to the historical and social background of the film, in that it could have been played in 1900.[12]
(2) Emotion and atmosphere. Classical music in *Picnic* follows an orthodox musical function in films.
(3) Cultural signification—signifying Patrician upper class and "European-ness," as a sign of British colonization and the "taming" of Australia.

However, the use of classical music in *Picnic* was subject to criticism by figures such as Bruce Smeaton, composer of the film's incidental music, who declared his disapproval of it.[13] Jack Clancy claims that viewers were puzzled as to why *Picnic* used classical music.[14] It is often considered a lazy way of adding music to film, working without specificity or innovation. For example, in the last scene of *Picnic*, Beethoven's "Emperor" concerto acts as fundamental background music. It functions as a general sonic supplement, supplying emotional continuity; yet it provides no tangible extra semantic information. This is to say, were one to employ a different piece of classical music in this scene, the effectiveness would most likely be comparable.

The discussion concerning the general function of classical music may be understood by adopting a *Gestaltist* approach (also sometimes called "Configurationism") such as attending to isomorphic aspects, the Law of *Prägnanz* and Law of Closure. As Max Wertheimer notes, "[t]here are wholes, the behaviour of which is not determined by that of their individual elements, but ... by the intrinsic nature of the whole."[15] The core of *Gestalt* psychology is that organisms are apt to be perceived as integrated configurations rather than disassembled pieces. Based upon this theory, classical music used in *Picnic* refers to its holistic elegant sense, rather than exploiting the specific musical peculiarities of different pieces of classical music. Details are almost meaningless. Classical music in *Picnic* appears to be a plural mixture from the music pantheon, functioning as a representative of high culture, in addition to empowering the film with a general sense of grace and classicism. In the scene represented in Figure 12.1, for example, the diegetic classical music establishes the sonic-realism and enhances the elegant atmosphere of the outdoor gathering. Without exploring the complicated background connotations of specific pieces of classical music, this scene produces a comprehensive effect

Figure 12.1 Diegetic classical music: the string quartet contrasts with the background landscape, with the pot plants demonstrating a mastery of cultivation in *Picnic at Hanging Rock* (British Empire Films Australia, 1975).

via the prominent general concept of "classical music." Furthermore, the string quartet performance in this scene drastically conflicts with the background landscape; instead of merging together, the "nature" and the "culture" exist separately and emerge as a bizarre juxtaposition. As Ross Gibson notes, the Australian landscape is not like "a nurturing mother or a placid locale for the arbitrary organization of social life,"[16] instead, nature and culture appear to be in conflict in *Picnic*.

Picnic is sometimes positioned as an effective example of Australian Gothic film. Gothic film is normally not assigned its own position as a genre,[17] as there are only "Gothic images and Gothic plots and Gothic characters and even Gothic styles within film, [… which] usually fall into the broader category of *horror*."[18] Xavier Aldana Reyes notes that the Gothic sets up "atmospheres of gloom and unease that may also play with shadows to create a pervasive sense of threat. It is also highly psychological and preoccupied with hallucinations, vivid dreamscapes (often nightmares)."[19] As Reyes stresses, compared to visceral "horror," the Gothic in cinema is subtle. It is normally related to nightmarish emotions, bloodcurdling atmospheres, and haunted hallucinations, and all these elements can make the film contexts more ambiguous. Due to the semi-hidden expressive approach of Gothic cinema, this "genre" is embodied by enigmatic aesthetics, driven by questions and mysteries which serve often to establishing an inexplicable matrix. Furthermore, in Gothic cinema, according to Isabella van Elferen, film music tends to play the role of the "eloquent narrator of the invisible."[20] To complement the visually hidden information, sound maintains its capacity to enhance terror, horror, and the uncanny.

It may seem initially to be incongruous to use classical music in an Australian Gothic narrative. As Peter Brooker notes, the Gothic indicates an aesthetic style that contradicts mainstream and conventional fashions.[21] However, an extraordinarily gruesome effectiveness may be created via unique audio-visual combinations. Figures 12.2 and 12.3, for example, reveal one male character's hallucination, where one of the missing girls appears and then disappears in a cave. These images underscored by classical music (Beethoven's piano concerto no. 5), are imbued with a strong sense of the enigmatic and uncanny. This results in an unexpected oscillation between romance and terror within the overwhelmingly strange landscape. This is an example of the power of the "counterpoint" between the image and the sound that Michel Chion stresses, the idea indicating the state "… where sound and image would constitute two parallel and loosely connected tracks, neither wholly dependent on the other."[22] Although, in this case, the classical

Figures 12.2 and 12.3 A girl appears and disappears in a cave in *Picnic at Hanging Rock* (British Empire Films Australia, 1975).

Figures 12.2 and 12.3 (*Continued*)

music is moderate and undynamic, which has the potential to amplify the eerie sense of horror through the extreme juxtaposition between the sound and the Gothic image.

Figure 12.4 Sara, an orphan, sits on steps after discovering that she will be returned to the orphanage in *Picnic at Hanging Rock* (British Empire Films Australia, 1975).Susan Dermody and Elizabeth Jacka coined the term "Australian Gothic Cinema" to identify a group of films in the 1970s, amongst which *Picnic* is considered one of the most notable examples.[23] Apart from being remarkably uncanny and unfathomable (see the image style of Figure 12.4), the aesthetic spectrum of Australian Gothic elicits specific cultural connotations of the past.[24] While the Gothic more generally shows an interest in the effects of the past, the Australian version is highly distinctive. As Gerry Turcotte states, "the Gothic has dealt with fears and themes which are endemic in the colonial experience: isolation, entrapment, fear of pursuit and fear of the unknown."[25] It is worth noting that in discourses about colonialism and culture, nature and the landscape are often significant ideas, as the manipulation of indigenous landscape is an essential part of colonialization. Furthermore, the uniqueness of the landscape in Australia endows it with the character of, as Gibson notes, "a paradox—half-tamed, yet essentially untameable; conceding social subsistence, yet never allowing human dominance."[26] Accordingly, landscape is therefore a distinctive implement

Figure 12.4 Sara's corpse in the garden in *Picnic at Hanging Rock* (British Empire Films Australia, 1975).

for Australian Gothic cinema to epitomize or emphasize inscrutable local culture, which questions or opposes the orthodox white European-culture brought by colonialization. Classical music in *Picnic* thereby becomes a jarring sonic expressive force. Although classical music is often depicted as conventional and "safe" in films, utilizing classical music in Gothic horror, in addition to incorporating it with pristine and startling landscapes, is a rare employment.

Panpipe: the Strange Sonic Fascination that Underpins Exotic Landscape

Alongside the unsolved mystery that happens to these girls, *Picnic* exhibits a cryptic and fantastic visual style that showcases Australian landscape and nature. The idiosyncratic trees in Figure 12.5 and the grotesque reptile in Figure 12.6 both epitomize the uncanny topography of this film. However, the enigmatic parameters of *Picnic* rely perhaps more heavily on its music, with the sound of the panpipe reinforcing the uncanny pictorial style. In order to construct a mysterious atmosphere, the soundtrack of *Picnic* introduces several exotic musical elements. Marjorie D. Kibby discusses the idea of the "Great Australian Silence," which she states, "signifies not so much an absence of sound but the

172   *Haunted Soundtracks*

Figure 12.5 Disturbing landscape in *Picnic at Hanging Rock* (British Empire Films Australia, 1975).

Figure 12.6 Animals in the disturbing landscape in *Picnic at Hanging Rock* (British Empire Films Australia, 1975).

presence of a silence coloured by unfamiliar sounds."²⁷ She may be referring to diegetic sounds, but this also goes for music. Amongst the unfamiliar sounds in *Picnic*, the unusual instrument the panpipes (Flûte de Pan, also known as pan flute or syrinx) is the most prominent one. It is an ancient instrument (a multi-tube pipe from areas including Asia, Africa, South America, and non-western Europe), associated with several mythological or legendary ideas, such as the god Pan (arguably one of the *genius loci* that I mentioned above). In Greek mythology Pan is associated with nature and the rustic; his traditional haunts are the wild mountains, mysterious rocks, sunless caves, and uncanny woods.²⁸ Similarly, the story of *Picnic* can also be associated with the legend of the Pied Piper of Hamelin, in which the title character attracts children with his magic pipe and makes them disappear into a mountain. While this is usually depicted visually as a single pipe, the disappearance of the girls in *Picnic* appears obliquely related to the mythological and legendary associations of the musical pipe, which marshals the film's soundtrack. The panpipe music in *Picnic* is derived from two traditional pieces—*Doina Lui Petru Unc* (n.d.) as well as *Doina: Sus Pe Culmea Dealului* (n.d.) recorded by Romanian panpipe musician Gheorghe Zamfir.²⁹

Ancient instruments or instruments from distant areas have the potential to appear enigmatic and perhaps even the incomprehensible for audiences unfamiliar with them. The spatially remote and antique sound of the panpipe corresponds with the senses of "isolation," "alienation," and "mystery," and this instrument's sound appears in accordance with girls' disappearance in the deep mountains in *Picnic*. Here, the panpipe stands in for the exotic. While this strange instrumental sound appears to function as an embodiment of the film's central enigma, with this music appearing at moments when the film emphasizes the inexplicable events, it is the melancholic tonality of the panpipe that exaggerates their unfathomability. To intensify its effectiveness, the panpipe in *Picnic* is recorded with a very marked reverb effect. This reverberation engenders the sense of larger space, further reinforcing the ethereal nature of the music. Peter Doyle elaborates on echo and reverb as:

> something to do with the supernatural … Sometimes dark, subterranean spaces were evoked; at other times this listener was put in mind of grand mountains and canyons. … [with] Echo and reverberation made it seem as though the music was coming from a somewhere—from inside an enclosed architectural or natural space or "out of" a specific geographical location …³⁰

Doyle's description could almost have been written with *Picnic* in mind. The reverberant sound enhances the tones of the panpipe, emphasizing slight changes in tone and breath. This panpipe music accords with the categories of "exotic music" and "world music." The former is associated with the idea of exoticism, which often relates to places that are unfamiliar to mainstream culture. According to Ceroni, Ma, and Ewerth, "Exoticism is the charm of the unfamiliar, it often means unusual, mystery, and it can evoke the atmosphere of remote lands."[31] The remote lands described here correspond to the outback of Australia, which can also be associated with the colonial anxiety of the landscape of this territory. As Johnny Milner notes, "the unknowable and expansive Australian outback ... with its droughts, floods, bushfires, tropical cyclones and dust storms—was completely out of sync with the relatively safe, cultivated and controlled European environment."[32] *Picnic* builds a sense of cultural and psychological instability for the people in their relationship with the unwelcoming Australian landscape, and particularly the girls. Exoticism in music is defined by Timothy D. Taylor as the "manifestations of an awareness of racial, ethnic, and cultural Others captured in sound."[33] Therefore, musical exoticism might be understood as based upon colonialism, imperialism, and globalization, affecting non-western Others.[34] In *Picnic*, the panpipe, as non-mainstream exotic music, supplies the audience with novel musical patterns along with exotic cultural associations, as part of the process of enabling an indistinct trace memory of the colonial past. The distinctive soundscape echoes the austere landscape with its spiritual connotations, whilst also suggesting anxiety and bewildering emotions. For instance, when four girls explore Hanging Rock before disappearing, the background panpipe music appears in accordance with the bizarre and distinctive landscape of Hanging Rock and the subsequent occult mystery.

It is striking that the panpipe, which is so prominent and memorable in the film, is not a traditional instrument from Australia, whereas *Picnic* has arguably become one of the most iconic expressions of the land of Australia. Perhaps, *this* partially reveals a confusion at the heart of the Australian identity; why utilize the panpipe in a seemingly incongruous territorial context? We could again approach this question via *Gestalt* aesthetics. Due to the exotic music's property of being obscure and unfamiliar, the audience will perceive a general sense of the idea of "enigma" rather than appreciating the musical ideas. The music communicates strangeness directly. In other words, rather than simply registering as music, it instead seems to pose a question. In *Picnic*, the panpipe tends to produce an overall aura of enigma. The fundamental logic here is to convey the general senses of "exotic" and "weirdness," by means of heterogeneous musical exotic undertones, which overwhelm the audience's ability to appreciate it as music.

Panpipe music could be categorized as "exotic music" or perhaps even "world music." The formal term of "world music" was coined in 1987.[35] Jocelyne Guilbault defines world music as music coming "from outside the 'normal' Anglo-American (including Canadian and Australian) … either from third world countries (Africa and the African diasporas), or from disadvantaged population groups in a general sense."[36] This musical idea has a tacit implication of cultural hegemony, and perhaps neo-colonialism, and is what John Connell and Chris Gibson define as "music tourism."[37] As Taylor notes, "world music … has some discernible connection to the timeless, the ancient, the primal, the pure, the chthonic; that is what they [the Western world] want to buy, since their own world is often conceived as ephemeral, new, artificial, and corrupt."[38] This is to say, what world music caters to is western consumers' curiosity about ancient and authentic music, stemming from non-western terrains. This is partially how the panpipe music in *Picnic* functions through its exotic unfamiliarity to a perceived mainstream of music, film culture and audience.

The seeming "authenticity" of world music is an attraction for western audiences. However, the concept of authenticity here is tricky in a sense, and perhaps is more "pseudo-authenticity." Connell and Gibson argue that the supposedly "authentic," with a genuine and uncontrived sense of "emotionality" is thought of as being absent in western music but evident in folk music from far-off places.[39] The supposedly authentic here indicates an "emotional authenticity" rather than a "factual authenticity," and thus the critical point is that it prompts listeners to "feel authentic" rather than the music needing to "be authentic." In this case, "world music" is not packaged for western audiences to appreciate and comprehend it appropriately (in its cultural context) but instead it is framed for romanticizing and fetishizing the notion of the "alien" and satisfying fetishism toward the unfamiliar music that exists in consumers' imaginations and consumer fascinations. Drawing upon Benedict Anderson's idea of imagined community,[40] Johnny Milner develops an idea called "imagined soundtracks" when referring to Australian cinema, which are "formulated and reformulated cultural inventions that seek to convey a sense of community and nationhood in the consciousness of the listener."[41] A similar discourse surrounding Australian sound tracks was instituted by Rebecca Coyle, who declares that the authenticity and hallmarks of film music in Australian cinema indicate the identity and ethnicity of this country.⁴² Yet such "imagined soundtracks" might echo imaginary nationalism and fabricated authenticity, particularly with respect to *Picnic* due to the incongruity of a non-native instrument such as the panpipe.

The pseudo-authenticity of world music provides a reasonable explanation as to why the non-Australian instrument (panpipes) is able to exude appropriate charm and power in this Australian film. Like the classical music in *Picnic*, panpipe music in this film also follows a Gestaltist approach, supplying an overall rather than a specific impression, to radiate a broad sense of the exotic in *Picnic*. In cinema, listeners perceive the musicality in a general and integral framework, and the cultural connotations or social background of film music may be negligible. This corresponds to the "Law of *Prägnanz*" in Gestalt psychology. *Prägnanz* is a German word that implies "pithiness" and "simplification."[43] In the light of the audience's perceptual patterns, they are likely to receive the information in a simplified matrix, addressing a minimal core of the whole. In other words, the film music in *Picnic* takes advantage of the general idea of being "exotic," rather than being approached as specific music or specific culture. In *Picnic*, ideas of enigma converge with symbols of the Other, manifested by "exotic music" or "world music." This is a *Gestaltist* sonic aesthetics, regardless of concrete and idiographic musical insights. *Picnic* manipulates both the familiar (classical music) and the exotic (panpipes) precisely as general sonic ideas rather than primarily as music in itself.

Ideological Otherness—Feminism in Cinema and the Australian New Wave

Anahid Kassabian contends that film music carries and crystallizes socio-cultural embodiments and ideological parameters.[44] There are specific connections between exotic music and ideological exoticism in *Picnic*. For example, as I noted, it is a representative film of the Australian New Wave, and it also has significant emphasis on feminine aspects. In an influential article published in the same year as the film was released, Laura Mulvey theorized the concept of the "male gaze," where a male (in the film or implied) acts as the bearer of the look toward the woman on screen, while women embody *to-be-looked-at-ness*.[45] The teenage girls in *Picnic* are consistently being watched by male characters. They are the quintessential product of the patriarchal society of Victorian Australia. Impeccable cheeks, slender fingers, girlish figures, soft hair, silk stockings, these girls are portrayed as perfect maiden symbols. During the Victorian era from 1837 to 1901, a woman's responsibility was centered on domesticity.[46] In this case, the status of the female characters in *Picnic*, which is set in 1900, is subaltern and subordinate to patriarchal society, and very clearly the second sex, to use Simone de Beauvoir's term.

These schoolgirls' status is also manifested by their clothes. At the beginning of the film, they help each other to tie their corsets (see Figure 12.7), which symbolize their restriction. Clothing for women during that era had stringent rules.[47] Interestingly, this strict code of behavior is in opposition to the untamed natural landscape in *Picnic*. In this film, there is a dichotomy between nature and culture, between the natural and metropolitan, between the well-regulated and lawless, between cultured elegance and the uncontrollable outback, which engenders the ideological and spiritual conflicts that are embodied by these young schoolgirls. When these seemingly tame girls are embedded in the primordial landscape during their picnic, the contradiction between sophisticated civilization and visceral nature epitomizes an intense incongruity, which corresponds precisely to the unorthodox deployment of music in the film. As I already mentioned, the juxtaposition of classical music (an emblematic embodiment of sophisticated white culture and civilization) and the panpipe music (an exotic and primitive music form) embodies a remarkable acoustic opposition and collision.

Figure 12.7 The girls fastening each other's dresses in *Picnic at Hanging Rock* (British Empire Films Australia, 1975).

Furthermore, the film portrays a homoerotic tendency in these girls, reinforcing the subordinate position of these female figures. With women being a subordinate "Other" during the era depicted, lesbianism was socially unacceptable and they therefore would have constituted a minority within the subjugated minority, the Other among the Other. These schoolgirls, with their intimations of being lesbians, symbolize the subordinate identities during that era, and the exotic music in this film similarly "contradicts" the dominant procedures of mainstream soundtracks and is able to sonically signify the ideas of the minority and exoticism.

The exoticism of the music dovetails with the Otherness of the female characters, and also with the stylings of the Australian New Wave. The Australian New Wave is considered to be "the search for national identity and the need for Australian culture to assert itself against perceived cultural imperialism, from both America and Britain."[48] Since the Australian New Wave aimed to diverge from the mainstream international film market, the music it applies could appear atypical. Furthermore, landscape could also be exploited as a significant hallmark that differentiated it from mainstream or overseas cinema, wherein distinctive Australian nature and topography counterpose with "civilized" white-colonial-based culture. To redouble the effect, film music could endow the featured landscape with a bizarre and inscrutable atmosphere, and this is nowhere more evident than in *Picnic*.

Conclusion

In *Picnic*, the soundtrack has a crucial but ambiguous presence, with its highly significant role in marshaling the film's ideas added to its underlining and embodying the film's central mystery. The correlation between *Picnic*'s thematic notions and music is between enigmatic ideas and "exotic music." Film music with exotic components is successful in fulfilling non-mainstream values, beliefs, and ideologies, which are embedded and engraved in the idiosyncratic Australian landscape topographically. In *Picnic*, in order to convey non-mainstream ideas such as women's position in Victorian Australian society or the stylings of the Australian New Wave, exotic musical forms like the panpipe are potent. In contrast, the classical music is an unambiguous symbol of white British colonial culture. However, the unexpected combination between the Australian Gothic images and the classical music in *Picnic* renders the conventional music strange and exotic. All these musical elements help stretch unfathomable ideas, deepening the intertextuality between the sonic novelty and the conceptual obscurity.

The schoolgirls are a site of competition between colonial culture and the "native," embodied by European classical music and the exotic panpipe music respectively. Although connected by the former, they are affected by the latter. Furthermore, the girls at Hanging Rock appear connected by implication with a structural absence: Australian aboriginal people. The "mystery" of the brutalization of the indigenous population by British colonization potentially indicates that Aboriginal people's absence in *Picnic* involves a substitution by the Caucasian girls. While the solution to the mystery (and the Aborigines) is absent, the panpipe music makes for itself an even stronger presence, as a spectral object that haunts the film with an insistent character yet uncertain importance. The panpipe music manifests a question and makes present the mystery itself.

At the end of this chapter, it is worth pointing out the Law of Closure here. As a *Gestaltist* approach, the Law of Closure tells us something of the way we recognize and navigate the film. If we consider *Picnic* as a "whole," there must be something significant hidden in the dark. In *Picnic*, although it is true that eccentric incidents happen, the film does not reveal the full picture of the events and narrative, and audiences are apt to supplement their puzzlement through macrocosmic imaginings. We cannot take the film "as is" due to its gaps in information and explanation, instead we are encouraged to address the enigma directly. The enigma is a structured absence or void where something is not necessarily visible, but nonetheless appears knowable. In this case, the panpipe music occupies this empty enigmatic space, coming to embody the film's mystery.

Notes

1   More information about Australian New Wave, see Brian McFarlane and Geoff Mayer, *New Australian Cinema: Sources and Parallels in American and British Film* (Cambridge: Cambridge University Press, 1992); Michael Walsh, "Building a New Wave: Australian Films and the American Market," *Film Criticism* 25, no.2 (2000): 21–39.

2   See films such as *Picnic*, *Walkabout* (Nicolas Roeg, 1971), and *Rabbit-Proof Fence* (Phillip Noyce, 2002).

3   Peter Pierce, *The Country of Lost Children: An Australian Anxiety* (Cambridge: Cambridge University Press, 1999), xii.

4   Interview with Peter Weir (published April 8, 2015), https://www.youtube.com/watch?v=9bHxj3QXqiY (accessed September 16, 2021).

5   Jonathan Rayner, *The Films of Peter Weir* (London: Cassell, 1998), 56.

6   Scholars such as Suzie Gibson and Saviour Catania, both admit the ambiguity of the narrative of *Picnic*.

Suzie Gibson, "The Embrace of Ambiguity in Joan Lindsay's *Picnic at Hanging Rock* and Henry James's *The Turn of the Screw*," *Antipodes* 33, no.1 (2019): 8–20; Saviour Catania, "The Hanging Rock Piper: Weir, Lindsay, and the Spectral Fluidity of Nothing," Literature/Film Quarterly, 40.2 (2012), 84–95.

In addition, in the book *Dreams within a Dream*, Peter Weir admits that in his films, there are normally no satisfactory resolutions, whereby audience's imagination tends to be provoked.

Michael Bliss, *Dreams within a Dream: The Films of Peter Weir* (Carbondale; Edwardsville: Southern Illinois University Press, 2000), 188.

7   Gibson, "The Embrace of Ambiguity in Joan Lindsay's *Picnic at Hanging Rock* and Henry James's *The Turn of the Screw*," 18.

8   Bruce Johnson and Gaye Poole, "Sound and Author/Auteurship: Music in the Films of Peter Weir," in *Screen Scores: Studies in Contemporary Australian Film Music*, ed. Rebecca Coyle (Sydney: Australian Film Television & Radio School, 1998), 129.

9   Music in Australia Knowledge Base has organized the limited research concerning the soundtracks of Australian cinema. See "Australian Feature Film Music in the Sound Era," Music in Australia Knowledge Base (published January 1, 2008), https://www.musicinaustralia.org.au/australian-feature-film-music-in-the-sound-era/ (accessed August 25, 2021).

10 Being accorded unanimous praise, the film music of *Picnic*, I suggest, balances the traditional dichotomy between commercial needs and aesthetic pursuit simultaneously. As an Australian film, it captures the western or international audiences' fetishism toward exotic unfamiliarity, without neglecting aesthetic values. In 2017 the soundtrack of *Picnic at Hanging Rock* was ranked twelfth among the 100 best film scores on a poll by ABC. See "Music in the Movies," ABC Classic FM, May 30, 2017, http://www.abc.net.au/classic/classic100/movies/#all (accessed October 7, 2021).

11 Theodor Adorno and Hanns Eisler, *Composing for the Films* (London: The Athlone Press, 1994), 15–16.

12 For more information as to the history of western music, see J. Peter Burkholder, Donald Jay Grout and Claude V. Palisca, *A History of Western Music* (New York; London: W. W. Norton & Company, 2014).
13 Conversation between Bruce Smeaton and the writer February 6, 1992. Jack Clancy, "Music in the Films of Peter Weir," *Journal of Australian Studies* 18, no.41 (1994): 29.
14 Ibid., 29.
15 Max Wertheimer, "Gestalt Theory," in *A Source Book of Gestalt Psychology*, ed. W. D. Ellis (New York: Kegan Paul, Trench, Trubner & Company, 1938), 2.
16 Ross Gibson, *South of the West: Postcolonialism and the Narrative Construction of Australia* (Bloomington; Indianapolis: Indiana University Press, 1992), 69.
17 As to the origin of the word "Gothic," as Peter Brooker declares, "with the pejorative connotations of uncouth, ugly, archaic or barbaric, derived from this association, …. It came thence to denote anything vast, gloomy or medieval in architecture or to describe signs of decay and wildness in buildings or landscapes." Peter Brooker, *A Glossary of Cultural Theory* (Oxford: Oxford University Press, 2003), 114–15.
18 Misha Kavka, "The Gothic on Screen," in *The Cambridge Companion to Gothic Fiction*, ed. Jerrold E. Hogle (Cambridge: Cambridge University Press, 2002), 209.
19 Xavier Aldana Reyes, *Gothic Cinema* (London; New York: Routledge, 2020), 8. For more information concerning the difference between the Gothic and the horror, see Isabella van Elferen, *Gothic Music: The Sounds of the Uncanny* (Cardiff: University of Wales Press, 2012), 36.
20 Ibid., 37.
21 Brooker, *A Glossary of Cultural Theory*, 115.
22 Michel Chion, *Audio-Vision: Sound on Screen* (New York: Columbia University Press), 35.
23 Jane Stadler, "Atopian Landscapes: Gothic Tropes in Australian Cinema," in *A Companion to Australian Cinema*, ed. Felicity Collins, Jane Landman and Susan Bye (New York: Wiley Blackwell, 2019), 337.
    For more information see Susan Dermody and Elizabeth Jacka, *The Screening of Australia Vol II* (Sydney: Currency Press, 1988); David Thomas and Garry Gillard, "Threads of Resemblance in New Australian Gothic Cinema," *Metro* 136 (2003): 36–44.
24 Stadler, "Atopian Landscapes," 336.
    For more information see Jonathan Rayner, "Gothic Definitions: The New Australian Cinema of Horrors," *Antipodes* 25, no.1 (2011): 91–7; Kay Schaffer, *Women and the Bush: Forces of Desire in the Australian Cultural Tradition* (New York: Cambridge University Press, 1988).
25 Stadler, "Atopian Landscapes," 336.

For more information see Rayner, "Gothic Definitions," 91–7; Schaffer, *Women and the Bush*.

26 Gibson, *South of the West*, 67.
27 Marjorie D. Kibby, "Sounds of Australia in Rabbit-Proof Fence," in *Reel Tracks: Australian Feature Film Music and Cultural Identity*, ed. R. Coyle (Sydney: John Libbey/Perfect Beat, 2005), 151.
28 Donald Barrett, "The Mythology of Pan and Picnic at Hanging Rock," *Southerly* 42, no.3 (1982): 300.
29 Gheorghe Zamfir's debut in film music area is a French comedy film *Le Grand Blond avec une Chaussure Noire* (Yves Robert, 1972).
30 Peter Doyle, *Echo and Reverb: Fabricating Space in Popular Music Recording, 1900–1960* (Middletown, Conn.: Wesleyan University Press, 2005), 5.
31 Andrea Ceroni, Chenyang Ma and Ralph Ewerth, "Mining Exoticism from Visual Content with Fusion-Based Deep Neural Networks," *Proceedings of the 2018 ACM on International Conference on Multimedia Retrieval* 18 (2018): 37.
32 Johnny Milner, "Australian Gothic Soundscapes: *The Proposition*," *Media International Australia* 148, no.1 (2013): 96.
33 Timothy D. Taylor, *Beyond Exoticism: Western Music and the World* (Durham; London: Duke University Press, 2007), 2.
34 Ibid., 1.
35 John Connell and Chris Gibson, "World Music: Deterritorializing Place and Identity," *Progress in Human Geography* 28, no.3 (2004): 342.
36 Jocelyne Guilbault, "World Music," in *The Cambridge Companion to Pop and Rock*, ed. Simon Frith, Will Straw and John Street (Cambridge: Cambridge University Press, 2001), 176.
37 Connell and Gibson, "World Music," 345.
38 Timothy D. Taylor, *Global Pop: World Music, World Markets* (London; New York: Routledge, 2014), 26.
39 Connell and Gibson, "World Music," 344.
40 Benedict Anderson, *Imagined Communities: Reflections on the Origin and Spread of Nationalism* (London: Verso Books, 2006).
41 Johnny Milner, "Sounding Country: Tracking Cultural Representations in the Soundtracks of Contemporary Australian Landscape Cinema," Ph.D. thesis (The Australian National University, 2016), 54.
42 Rebecca Coyle, ed., *Screen Scores: Studies in Contemporary Australian Film Music* (North Ryde: The Australian Film, Television and Radio School, 1998), 10–11; for more information, see Rebecca Coyle, ed., *Reel Tracks: Australian Feature Film Music and Cultural Identities* (Sydney: John Libbey Publishing, 2005).
43 Kurt Koffka, *Principles of Gestalt Psychology* (London: Kegan Paul, Trench, Trübner & Co., 1936), 138.

44 Anahid Kassabian, *Hearing Film: Tracking Identifications in Contemporary Hollywood Film Music* (New York; London: Routledge, 2001), 29.
45 Laura Mulvey, "Visual Pleasure and Narrative Cinema," *Screen* 16, no.3 (1975): 11.
46 Lynn Abrams, "Ideals of Womanhood in Victorian Britain," *BBC History* 9 (2001): 1.
47 For supplementing, as Lynn Abrams notes, "the nineteenth century women's fashions became more sexual—the hips, buttocks and breasts were exaggerated with crinolines, hoopskirts and corsets which nipped in the waist and thrust out the breasts. ... women became walking symbols of their social function—wife, mother, domestic manager"; "The fashion for constricting corsets and large skirt served to ... the physical constraints on her activities."
Abrams, "Ideals of womanhood in Victorian Britain," 3.
48 Rochelle Siemienowicz, "Globalisation and Home Values in New Australian Cinema," *Journal of Australian Studies* 23, no.63 (1999): 49.

13

Outside Inside: Nature, Gender, and the Altered Domestic Space in *Possum* (1997) and *Nature's Way* (2006)

Andrea Wright

The landscape and nature are central to New Zealand culture and cinema, and screen representations have been influenced variously by a dual conception of landscape identified by Claudia Bell as both beguiling and perilous.[1] Bell describes a peculiarly bifurcated landscape experienced by settlers:

> There are two versions of romanticized landscape. Landscape is either beautiful but potentially dangerous: sanctified, visited, enjoyed, photographed, then left; a vision to inspire. Or it is beautiful and beautifully cultivated, a tribute to both nature itself and to the efforts of human labour.[2]

This definition has become pivotal to the Kiwi Gothic and its unsettling and sometimes contradictory imaginings of the natural environment and settler relationships with it. Short films *Possum* (Brad McGann, 1997) and *Nature's Way* (Jane Shearer, 2006) come from this tradition. Both present a troubling vision of human interaction with nature. As Ian Conrich has observed, "[t]he binary opposition of domestic/wild is central to many examples of Kiwi Gothic in which the home of the settler offers shelter against the forces of the wilderness."[3] These films upset the supposed security of such binaries and explore a blurring of boundaries as the mysterious forces of nature encroach upon the domestic space. Cinematically the films, although stylistically different, visually and aurally capture the strangeness of a natural environment that remains outside human control. The soundscapes are especially pronounced and render the environment eerie and threatening. The films also situate men in a particularly uneasy association with the natural, which disrupts settler mythologies of man's mastery of nature. Aligned with the feminine, nature cannot be contained by masculine action and the domestic spaces, which are, unusually, masculinized, become susceptible to its strange power. This chapter will argue that these films dramatize the theme of outside inside and deliberately unsettle relationships between nature, gender, and the domestic space.

The importance of the short film is documented by Trisha Dunleavy and Hester Joyce in their consideration of the New Zealand film and television industries. They note that short fiction films were seen as progressively important in raising the profile of New Zealand film, especially after a revision of the New Zealand Film Commission's Short Film Fund remit in 1995. The Fund, originally set up in 1986, provided training opportunities through small investment and helped to foster emerging talent. In 1995, there was an increased focus on developing a progression into features and in building an international reputation and recognition.[4] From the late 1980s, New Zealand shorts garnered considerable praise and critical acknowledgment in international festivals and competition, thus contributing to the visibility and viability of the national industry. The long-term result has been a rich and varied body of films that are as distinctive and significant as the country's feature output. Thematically, visually, and sonically, there are many parallels between New Zealand shorts and feature films. Those imbued with a sense of dark foreboding form a recognizable strand of the national output, and, along with an equally prevalent literary cluster, have been labeled variously as Antipodean, New Zealand, or Kiwi Gothic. A distinctly New Zealand Gothic has been linked to the formation of nation and the attendant settler anxieties. Conrich notes that "Kiwi Gothic films are marked by being repeatedly set in fractured neighbourhoods and dwellings, not just small towns and cities, but commonly in homes and houses, which are sites of invasion from foreign, inhuman or supernatural forces. The extent to which this occurs within Pakeha texts suggests a local unsettlement or vulnerability in reaction to a sense of space."[5] Feelings of dislocation and susceptibility are particular to *Pakeha* (white New Zealanders of predominantly European decent) narratives and sites of refuge, especially the home, are prone to a greater and more mysterious power that lurks beneath the surface of "normality." The sense of being uneasy and exposed is also a consequence of, as Alfio Leotta points out, "the settler anxiety that derived from the confrontation with a hostile and alien environment, such as the native New Zealand Bush."[6] It is a landscape of otherness that is resistant to cultivation and civilization, and, even when it is settled, there remains a sense of precariousness and uncertainty.

William J. Schafer, recognizing the significance of Gothic literature and film, asserts that there is a particular "necessity of ghosts" in Aotearoa New Zealand. He argues that a "hauntedness of the landscape" and a persistent feeling that it is a "land of unseen forces, of imminent (and immanent) threat, of undead or revenant spirits" provides a common link between Māori and *Pakeha*.[7] Crucially this is part of the nation's becoming:

> In the process of self-definition, cultures need to pass through a stage of hauntedness. This may be a simple analogy for the birth of historical consciousness—one way to gain historical rootedness in other than an abstract, intellectual way is to feel that the past is a horror waiting to reinvade the present. If you feel raw, young, unformed, lacking in historical status, a way to gain stature is to acquire suitably ancient ghosts.[8]

In young nations lacking ancient foundations and ancestries, Gothic fictions function as a "disguised history."[9] For New Zealand, the landscape and nature manifest archaic terrors in the absence of architectural sites of cultural heritage that could harbor ancient spirits. This is tied to complex feelings of home and belonging, the homely and unhomely, as Schafer puts it, "to possess a place is to be possessed by it."[10]

The Freudian sense of the *unheimlich* or the uncanny is frequently mobilized to define the colonial and postcolonial experience of space and place. The uncanny is defined by Nicolas Royle as a crisis of the proper and a crisis of the natural. He argues that the uncanny is not "simply an experience of strangeness and alienation. More specifically it is a peculiar comingling of the familiar and unfamiliar. It can take the form of something familiar unexpectedly arising in a strange and unfamiliar context, or of something strange and unfamiliar unexpectedly arising in a familiar context."[11] The process of settlement itself forces the familiar into and onto the unfamiliar and attempts to reconcile them are physically, psychologically, and ideologically turbulent. Importantly, Lizabeth Paravisini-Gebert writes that the colonial space is "by its very nature a bifurcated, ambivalent space, where the familiar and unfamiliar mingle in an uneasy truce."[12]

It is this notion of an "uneasy truce" between the inside and the outside that impacts upon *Possum* and *Nature's Way* and precipitates a mingling of the familiar and the unfamiliar and a seepage between interior and exterior within the texts. Moreover, both are imbued visually and sonically with a hauntedness that hints at unseen forces and looming dangers that are uncontrollable. *Possum* revolves around a settler family, a widowed trapper, and his three children, who live in isolation in the bush. The film is narrated from the subject position of the middle child, Little Man (Martin Taylor). The youngest child, a feral girl, Kid (Eve-Marie Brandish), is killed by a trap set by her father (Stephen Papps). *Nature's Way* focuses on an unnamed child abductor and killer (Matthew Sunderland) who is mysteriously encouraged back to the scene of his crime where he is punished by the natural/supernatural.

In both films nature itself if imposing, overwhelming and, paradoxically, despite its scale, claustrophobic. From the opening shots, the sense that the environment is confining the characters is prevalent. *Possum* begins with a black screen, and the sounds of rushing wind and a squawking bird are heard before an image of the edge of a thick forest fades in. The density of the trees creates the impression of a dark impenetrable place cordoned by the tall bar-like trunks. An enclosed wooded area is revealed as the scene changes to the trapper and his son making their way to retrieve their kill. The persistent swoosh of the wind and frequent bursts of birdsong are interrupted by loud rustling and cracking as they make their way through the pine-needle covered bush. A close-up of a large dead hare confirms the discord between man and nature and foreshadows the tragic events to come. As the boy looks up into the canopy above him a single high-pitched non-diegetic note sounds, and he whispers: "Sometime a possum get caught too." His softly spoken voiceover throughout the film is a raspy, barely audible whisper that infers telling of secrets and hints that he understands the imposition of human presence.

Maud Ceuterick identifies that *Possum* has a "porous environment" that blurs the borders of the human and animal world and sound especially plays a significant role in creating this porousness.[13] As she observes, the atmospheric sounds of the forest are constantly audible on the soundtrack, although in places they are interrupted and obscured with loud and abrasive sounds produced by the human characters—raised voices, abrupt and aggressive bangs and crashes and the persistent hum of a finger circling the rim of a drinking glass.[14] Furthermore, the environmental noise, non-diegetic instruments, and Foley connect and overlap, thus obscuring their origin and disorientating the viewer. The loudest sounds are often produced by Kid as she imitates the wild animals she sees on the pages of an encyclopedia. These various noises, Ceuterick argues, place the film, like Alison Maclean's 1989 *Kitchen Sink*, at "the border between magic realism and horror by investing the domestic with wildness and animality."[15] Maclean's almost Lynchian black-and-white short is a nightmarish tale of a woman who pulls at a long hair caught in a plughole to reveal a tiny hair-covered human-like creature. The alien form that invades the home rapidly grows into a full-sized man which she shaves and dresses. Metallic, almost industrial, reverberations are blended with an undefined atmosphere hum and birdsong on the soundtrack. Intermittently, sounds of water running and draining, the unpleasant noise of suction in a blocked drain, and a razor scratching at skin and hair intensify the oppressive, often repulsive images. As a result, in both examples, the domicile and what lies beyond are no longer distinct, and the sanctuary of home is permeable. In *Possum*, especially, "sounds convert the house into a porous space with the adjacent forest. Animality seeps through the walls and into the characters, emphasizing the animal side of human beings."[16] Consistently, domestic routines are disrupted by the natures' sonic presence. For instance, as he bathes his sister in a tin tub in the dwelling, Little Man whispers: "Kid got power of the howling wind." A rushing and humming become louder on the soundtrack and is punctuated with an irregular single piano note, a common feature of Tom Bailey's score throughout the film. The image briefly changes from a close-up of Kid to a low-angle tracking shot moving rapidly, animal-like, through the bush, and then back again.

Visually this collapsing of boundaries is further supported by color and framing. The sepia tint of the film, used, according to director, Brad McGann, to make it timeless and ambiguous,[17] further obscures the tonal distinction of inside and outside. The forest and the home, due to the brown hues and tight angles, are both shadowy and oppressive. Specifically, as Leotta notes, "the film is based on the opposition between two distinct and yet oddly similar cinematic spaces: the forest and the family home."[18] The home itself is not welcoming or comfortable, instead it is sparse and rudimentary, and a functional masculine rather than feminine space. The bush and the home are strange and forbidding and both contain secrets.

The opening of *Nature's Way* similarly indicates the oppressiveness of nature and from the title sequence, a low hum permeates the audio channel. The first shot is of rivulets of rain running down a windscreen. In the second, the repositioned camera gives a sense of the enclosed space of a vehicle with the silhouetted figure sitting in the driver's seat. The gentle patter of water and the mechanical squelch of the windscreen wipers create a monotonous rhythm. The scene then changes to capture the wider environment: angular roofs under a proportionately larger, grey, clouded sky and a lush, leafy forest. The "porous environment," although represented differently, is also apparent. The most dominant diegetic noises are rain, the wind, and the rustle of the trees, which provide a consistent ambience. Even inside the man's home, the sounds, especially, the quiet swish of the wind, are perceptible in the background. Aurally the outside bleeds subtly into the domestic space. Rachel Shearer's non-diegetic soundtrack is mournful and comprised of long, droning tones occasionally interrupted by staccato piano notes that accentuate moments of tension and signal foreboding. In places, the sounds intermingle indistinguishably with the environmental resonances. Human noise is largely abstruse, and the absence of audible dialogue alienating. The audience briefly hear the girl (Katrin O'Donnell) scream as she is pulled into the undergrowth, the man's breathless grunting after he commits the murder and later when he is pursued in the forest, and the distant disembodied cries of "Yvonne" as people search for the child. The most recognizably human sounds come from an unseen television set in the man's home.

Fred Botting has argued that sound in Gothic fiction is overlooked despite the fact that "the affective decoration of gothic fiction is intensified by its sonic chiaroscuro."[19] Writing about the work of Edgar Allen Poe, and in particular, *The Fall of the House of Usher*, he observes that "[u]ncanny reverberations are part of the tale's phantasmagoria of doublings and crossings between internal and external realms, perceptual and medial registers, imagination and reality in movements that never quite restore conventional boundaries between cause, effect, truth, convention or fiction."[20] The notion of intensification through sonic chiaroscuro can be applied to both texts discussed here, as can the sense of crossings and doublings that trouble the perception of boundaries. In *Nature's Way*, the visual collapsing of boundaries has consistencies with *Possum* in a number of ways. While not monochrome, it is subdued in tone with a predominance of dark greys, greens, and blues. Even daylight is muted and, in keeping with the notion of the judgment of nature, the natural environment is bearing down on the protagonist. Aurally the persistence of environmental sound prevents a distinction between interior and exterior.

Most poignantly, in *Nature's Way*, the man's feeling of unease and the sensation he is being observed by something unseen out in the woods beyond his fenced garden demonstrate the fragility of the supposed security of home and signal an eruption of unhomeliness. Although his neat lawn is cordoned, the wilderness is just outside the borders and the pervasiveness of natural acoustics is a reminder of its uncontainable power and presence. The possibility of invasion of the outside inside becomes literal as a shoot grows from under the house and penetrates the stark white and grey of the home's interior. The creeping plant is revealed to have wrapped itself around the pipework under the dwelling. As the man peers through the wooden slats beneath his house the shrill sound of crickets and the rattle of insects becomes piercing and is accentuated by a sharp non-diegetic tone. He is haunted by the specter of the girl and glimpses her form amid the swaying undergrowth, and the sounds of the bush bear down on him. The use of repetition in the sequences and sound is also important in underscoring the uncanny nature of the events in the film in, for example, the journeys the man makes to the forest on the winding, rain-soaked road, first to kill and then to return to the scene of his crime. His car, first a red and then a silver one, is parked and partially covered by the undergrowth. On two occasions, the man bursts out of the bush wide-eyed, panting, omitting intermittent grunting. The noise and the action are ambiguous—excitement, release, fear, guilt? As in *Possum*, the potential animality of humans is exposed and heightened by sound.

Both films, through their visual and sonic schemes, consistently hint at something beyond, a crisis of what is thought to be fixed and stable. Writing about German Expressionist Cinema, Paul Coates suggests that "[t]he world becomes uncanny when it is perceived as no longer simple substance, but also as shadow, a sign of the existence of a world beyond itself, which it is nevertheless unable fully to disclose."[21] *Possum* and *Nature's Way* hint at a world beyond what is simply perceived. Something exists at the periphery that leaks into the flimsy "normality" of the domestic space. Indeed, Freud's discussion of the proximity of the *heimlich* and the *unheimlich* highlights an unsettling conception of the home. In *Possum*, the fractured family are removed from civilization in their isolated environment. Inside the home a child is mistreated by her father and sister—she is fed like a dog under a table, she is relentlessly teased and as punishment is restrained. As Greg Dolgopolov points out there is also a "hint of incest"[22] between the father and eldest daughter, Missy (Alexia Verdonkschot). In *Nature's Way* the presumed ordinariness of a well-maintained home—indicated outwardly by its identical look to others in the suburban development, neatly trimmed lawns and clean car in the driveway—harbors a murderer.

Royle's notion that the uncanny is "a crisis of the proper and a crisis of the natural" resonates with the representation of gender within both films. The disruption of the symbolic order and patriarchal dominance at the metaphorical hands of nature is particularly interesting. In *Possum*, the troubled family unit, according to Dolgopolov, "articulates the contradictions between family expectations, paternal rule, approval, the power of books and freedom."[23] Patriarchal authority is cruel and brutal, and perhaps the seemingly most civilized character Missy, who wears a neat floral dress and makeup, is spiteful with a capacity for violence. As already noted, the film specifically questions the distinction between humanity and animality. As Cueterick notes, "[t]hey become porous terms as the spectator cannot draw a clear-cut line between the two and cannot tell who is the wildest of all the characters is, Kid, Dad or Missy."[24] Such confusion of what is human and animal is echoed by Dolgopolov who labels the family as a whole "feral."[25]

In *Nature's Way*, the man feigns normality by going through the motions of leading an ordinary life. He tends his house inside and out, and he appears to be settled with a female partner. However, as Adam P. Wadenius points out in his discussion of the 1998 film *Happiness*, the symbolic order maintains itself through borders, and the "deviant transgressions" of the monstrous male "most clearly points to the fragility of the symbolic order."[26] Moreover, the male monster "comes to signify the abject in his inability to perform proper masculinity."[27] Despite the mask of normality, *Nature's Way* director Jane Shearer underscores the horrific and abject, not only through the man's actions, but also in his visual and sonic representation. His face, when filmed in extreme close-up, distorts his features and shows imperfections on his skin and his unkempt hair. As he watches the children crossing in front of his car the rain, in between the swish of the windscreen wipers, clouds his face. After the murder his eyes are shown with heavy bags underneath, his hair damp with sweat and rain and he is omitting unpleasant animal-like groans. He becomes monstrous, to borrow Barbara Creed's definition, by not taking up proper gender roles (he kills a child rather than fulfilling a paternal role) and he potentially (although the film does not make explicit that he is a pedophile) exists at a border between normal and abnormal sexuality.[28] The knife that he has used to kill the girl is wiped on a piece of bread, and a full screen shot shows blood soaking into the white, textured, sponge-like surface. This is immediately followed by a short scene of the man eating a sandwich inside the car. Whether it is the blood-soaked bread is unclear, but the abject nature of his actions is implied by the visual montage, while the score's long echoey notes and the constant pitter-patter of rain accentuate the ambivalence by providing a sonic link between the sequences.

The central female children in both films exist in a state of liminality somewhere in between the realm of the ordinary and the supernatural. Kid is, to use McGann's description from the script, "a feral looking thing almost half animal."[29] Yvonne from *Nature's Way* appears to exist in a state between the living and the dead. The abstruse representation of the characters encourages the audience to perceive them as both abject and victims that demand sympathy. The representations of Kid oscillate between the demonic, wild creature biting and scratching and omitting shrieks and growls, and the pitiful child underneath the house where Little Man tells the audience the wild dogs come to lick water from the pipes. Kid is rendered visually and aurally abject, and the term, as defined by Julia Kristeva as having no respect for borders and rules, is particularly important.[30] This is achieved by her framing, her loud animal impersonations (which are rendered doubly uncanny because they are real animal sounds dubbed in post-production), and her resistance to patriarchal structures. Without language she remains outside the symbolic order. Little Man later throws away the makeshift wooden cross marking Kid's grave, thus also distancing her from the rituals of Christianity. He also leaves his bedroom window open at night in case she comes back, perhaps indicating that she cannot be contained, even in death. Kid remains in the realm of the semiotic with no shame and without boundaries or rules imposed by language. She is not governed by basic repression and has no concept of surplus repression. Yvonne, the child murdered in *Nature's Way*, is made abject by becoming a reanimated corpse. Moreover, the sickeningly loud crunch of deep roots breaking as the man lifts her long-dead body from the ground heightens the sense of horror. The child is literally absorbed into the environment and avenged by the power of a malevolent, it is reasonable to assume, Mother Nature.

Both films also articulate a fear of the failure of colonization and settlement. The absence of a secure, functioning family unit and the brittleness of patriarchal structures imposed by settlement are exposed. The representation of the adult males in both films, for instance, challenges the mythology of the stoic settler male. Both men are presented as performing traditional masculine identity, the father in *Possum* through his loud, brutal authority and the framing via low camera angles that make him appear large and imposing. The faux normality of the man in *Nature's Way* is accompanied by a *mise-en-scène* of domestic order. But neither are "proper" masculine figures. Misha Kavka explicitly reads *Possum* as a "story of settlement gone wrong."[31] Far from progressing to a state of stability through the establishment of the family, she argues that "[t]he three children in this settler family themselves represent an ironic reversal of civilized progression, in a kind of Darwinian devolution from youngest to oldest."[32] Not only is Kid a pest caught in the trap like a possum, people more generally are pests, thus making the film, according to Kavka, "a story of double non-belonging."[33]

The not belonging is also emphasized by the out-of-place-ness created by the visual and aural juxtaposition of signs of civilization in the untamed environment. In *Possum*, when their father is away the children look through mementoes belonging to their dead mother in a dusty jewelry box. The accompanying non-diegetic score plays ethereal, slightly off-key, music box chords, thus suggesting another type of haunting—the matriarch's ghostly aura. Missy picks up and plays with a string of pearls that are stark and bright in the gloom of the house. The shimmering surface and harsh rattle of the beads on the wooden floor are used to tease Kid as her sister taunts "here kitty kitty." When Kid takes the necklace from her sister they scuffle and the younger girl ferociously growls, shrieks, and bites in retaliation. As punishment, she is restrained and tied to a cot as she struggles and squeals. In *Nature's Way*, the child's oddly bright wellington boots and dropped plastic pen stand out against the dark green hues of the environment. Inside the home, the man's vacuum cleaner is loud and abrasive, thus rendering the common domestic sound an unwelcome intrusion. *Nature's Way*, although contemporary set in an established suburban environment, serves as a reminder that settlement encroaches upon nature and the boundary between civilization and the wild is permeable. As Conrich notes, "[h]is home is a subdivision, built on land cut from the native bush, with nature looming large on the edge of his property."[34] In both films, then, the noise and artifacts of human habitation are an imposition on the natural. This is a technique frequently mobilized by other New Zealand filmmakers such as Vincent Ward and Jane Campion to underscore the precariousness of settlement as unwelcome and incongruous.

The unease with belonging and home prevalent in New Zealand culture are symptomatic of settlement anxieties, and as Kavka observes, are tied up with an uncomfortable relationship with the past where memories of taming of the land and encounters with the original settlers continue to intrude the present. Specifically, Kavka, argues,

> [t]he result is a kind of ambivalent, permeable gothic domicile, both in the sense of the haunted house as home (to ghosts) and in the sense of New Zealand as home (to settlers). In this double meaning of gothic domicile cannot the material history of the land be said to haunt New Zealand stories as Old World spectres haunt a house?[35]

Home as residential building and home as the inhabited land both have the potential to harbor spirits, a prospect that might be amplified if the two intermingle and become less distinct. Accordingly, the domestic space in both *Possum* and *Nature's Way* is under threat from, haunted, and altered by, the presence of nature which is consistently signaled in the visual schemes and soundscapes. Far from establishing a clear demarcation between the inside and the outside, the outside seeps inside, thus upsetting traditional binaries. In these Gothic cinematic spaces, the home is not safe, it is a place of secrets and ghosts, and it is penetrable. Sound is especially important in muddling perceptions of interior and exterior and the subtle sonic reverberations of wind, rain, rustling leaves, and wildlife haunt the homes and their residents. Furthermore, the alignment of a mistreated and a murdered child with nature imbues them with mysterious energy that manifests in aural and visual signs that are ominous and inescapable. The films consistently challenge established notions of nature, gender, identity, and home through a dramatic interplay of outside and inside that is vividly rendered by imaginative visuals and creative sound design. The films make it possible to see and, significantly, hear the curious, eerie, and irrepressible power of nature in an antipodean setting.

Notes

1 Claudia Bell, *Inventing New Zealand: Everyday Myths of Pakeha Identity* (Auckland: Penguin Australia, 1996), 29.
2 Ibid.
3 Ian Conrich, "New Zealand Gothic," in *A New Companion to the Gothic*, ed. David Punter (Chichester: Wiley Blackwell, 2012), 403.
4 Trisha Dunleavy and Hester Joyce, *New Zealand Film and Television: Institution, Industry and Cultural Change* (Bristol and Chicago: Intellect, 2011), 146–7.
5 Conrich, "New Zealand Gothic," 402.
6 Alfio Leotta, "From Comic-Gothic to 'Splatstick': Black Humour in New Zealand Cinema," in *Directory of World Cinema: Australia and New Zealand*, ed. Ben Goldsmith and Geoff Lealand (Bristol and Chicago: Intellect, 2010), 296.
7 William J. Schafer, *Mapping the Godzone: A Primer on New Zealand Literature and Culture* (Honolulu: University of Hawai'i Press, 1998), 137.
8 Ibid., 138.
9 Ibid., 139.
10 Ibid., 142.
11 Nicolas Royle, *The Uncanny* (Manchester: Manchester University Press, 2003), 1.

12 Lizabeth Paravisini-Gebert, "Colonial and Post-Colonial Gothic: The Caribbean," in *The Cambridge Companion to Gothic Fiction*, ed. Jerrold E. Hogle (Cambridge: Cambridge University Press, 2002), 233.
13 Maud Ceuterick, "Porous rural spaces in Possum," *Short Film Studies* 6, no.1 (2016): 53.
14 Ibid.
15 Ibid., 54.
16 Ibid.
17 Richard Raskin, "An interview with Brad McGann on Possum," *Short Film Studies* 6, no.1 (2016): 39.
18 Alfio Leotta, "Possum's Cinematic Space: Landscape, Alienation and New Zealand Gothic," *Short Film Studies* 6, no.1 (2016): 46.
19 Fred Botting, "Poe, Voice and the Origin of Horror Fiction," in *Sound Effects: The Object Voice in Fiction*, ed. Jorge Sacido-Romero and Silvia Mieszkowski (Boston: Brill Rodopi, 2015), 73.
20 Ibid., 75.
21 Paul Coates, *The Gorgon's Gaze: German Cinema, Expressionism and the Image of Horror* (Cambridge: Cambridge University Press, 1991), 1.
22 Greg Dolgopolov, "*Possum*: The Fractured Family and Turning Animal," *Short Film Studies* 6, no.1 (2016): 58.
23 Ibid.
24 Ceuterick, "Porous Rural Spaces in Possum," 55.
25 Dolgopolov, "*Possum*," 58.
26 Adam P. Wadenius, "The Monstrous Masculine: Abjection and Todd Solondz's *Happiness*," *Horror Studies* 1, no.1 (2010): 135.
27 Ibid.
28 Barbara Creed, *The Monstrous-Feminine: Film, Feminism, Psychoanalysis* (New York and London: Routledge, 1993), 11.
29 Brad McGann, "The Original Screenplay: Possum," *Short Film Studies* 6, no.1 (2016): 27.
30 Julia Kristeva, *Powers of Horror: An Essay on Abjection* (New York: Columbia University Press, 1984), 4.
31 Misha Kavka, "The Settlement Trap," *Short Film Studies* 6, no.1 (2016): 69.
32 Ibid., 68.
33 Ibid., 69.
34 Conrich, "New Zealand Gothic," 403.
35 Misha Kavka, "Out of the Kitchen Sink," in *Gothic NZ: The Darker Side of Kiwi Culture*, ed. Misha Kavka, Jennifer Lawn and Mary Paul (Otago: Otago University Press, 2006), 57.

Contributors

Daniel Bishop teaches Musicology at the Jacobs School of Music, Indiana University, USA. His research interests include film music and the aesthetics of film sound, twentieth-century art music, and the history of sound reproduction.

Lisa Coulthard is Full Professor of Cinema and Media Studies in the Department of Theatre and Film at the University of British Columbia, Canada. She has published extensively on film violence, film sound, and film-philosophy. She currently holds a Social Science and Humanities Research Council of Canada (SSHRC) Insight Grant studying the fight scene in cinema and a SSHRC Insight Development Grant on Digital Dark Tourism. She is currently working on monographs on the fight scene in cinema and on sound and violence.

K. J. Donnelly is professor of Film and Film Music at the University of Southampton, UK. He is author of *The Shining* (2018), *Magical Musical Tour: Rock and Pop in Film Soundtracks* (2015), *Occult Aesthetics: Sound and Image Synchronization* (2013), *British Film Music and Film Musicals* (2007), *The Spectre of Sound* (2005), and *Pop Music in British Cinema* (2001); and editor of *Film Music: Critical Approaches* (2001), co-editor (with Phil Hayward) of *Music in Science Fiction Television: Tuning to the Future* (2012), co-editor (with Will Gibbons and Neil Lerner) of *Music in Video Games: Studying Play* (2014), co-edited with Ann-Kristin Wallengren, *Today's Sounds for Yesterday's Films: Making Music for Silent Cinema* (2016), co-edited with Steve Rawle, *Hitchcock and Herrmann: Partners in Suspense* (2017) and co-edited with Beth Carroll, *Contemporary Musical Films* (2018). He edits the "Music and the Moving Image" book series for Edinburgh University Press and "Palgrave Studies in Audio-Visual Culture" for Palgrave Macmillan, and is on the editorial boards of seven journals.

Jady Jiang is a Ph.D. student in the Film Department at the University of Southampton, UK. Her thesis addresses music and enigma in films, looking into how music can be used to create aesthetic confusion and a space for conceptual potential. She has presented papers at the Society for Cinema and

Media Studies (2022), the "Music and the Moving Image" conference at NYU (2020, 2023), and the British Association for Cinema and Media Studies (2023).

Danijela Kulezic-Wilson was a lecturer in Music at University College, Cork, Ireland. She studied musicology at the University of Belgrade, Serbia, and obtained her doctoral thesis in comparative study of music and film at the University of Ulster, UK. Danijela is the author of *Sound Design Is the New Score* (2019), *The Musicality of Narrative Film* (2015) and co-editor (with Liz Greene) of *The Palgrave Handbook of Sound Design and Music in Screen Media: Integrated Soundtracks* (2016).

John McGrath is a Lecturer in Music at the University of Surrey, UK. His research is concerned with how one format or artform can flow into and influence another. His monograph *Samuel Beckett, Repetition and Modern Music* (2018) explores the interactions and cross-pollination of music and literature while other recent publications investigate the transmedial work of Laurie Anderson and David Lynch. He is a committee member of the International Guitar Research Centre (IGRC) and is currently co-editing the collection *Twenty-First Century Guitar* for Bloomsbury.

Aimee Mollaghan is a Senior Lecturer in Film and Screen Studies at Queen's University, Belfast, UK. She is the author of *The Visual Music Film* (2015). Her research is grounded within music, sound, and the moving image. Her current scholarship is centered on the relationship between psychogeography and hauntology to the moving image, visual music, and television music. She continues to publish in these areas.

Paul Newland joined the University of Worcester, UK, in 2020 as Director of Research and Knowledge Exchange for the College of Arts, Humanities and Education. He previously worked at Bath Spa University and Aberystwyth University. He is the author of *British Films of the 1970s* (2013) and *The Cultural Construction of London's East End* (2008), and the editor of *British Rural Landscapes on Film* (2016) and *Don't Look Now: British Cinema in the 1970s* (2010). He is currently working on a jointly edited book, *British Art Cinema*.

Jamie Sexton is a Senior Lecturer in Film and Media at the University of Northumbria, UK. His publications include *Music, Sound and Multimedia: From the Live to the Virtual* (edited collection, 2007), *Experimental British*

Television (edited with Laura Mulvey, 2007), *Alternative Film Culture in Inter-War Britain* (2008), *Cult Cinema* (with Ernest Mathijs, 2011), *No Known Cure: The Comedy of Chris Morris* (edited with James Leggott, 2013), and *The Routledge Companion to Cult Cinema* (with Ernest Mathijs, 2018). He is currently writing a book called *Freak Scenes*.

Jessica Shine completed her Ph.D. at the University College, Cork, Ireland, and is a Lecturer in Creative Digital Media at Munster Technological University, Ireland. Her main areas of interest are soundscapes, aesthetics, and narrative, and she has published material about TV shows *Peaky Blinders* and *Sons of Anarchy*.

Craig Wallace completed a Ph.D. in the School of Arts, English and Languages at Queen's University Belfast, UK, in 2018. He is a teaching assistant in English, Film Studies, and Broadcast Production. He contributed to *Of Mud and Flame: A Penda's Fen Sourcebook* (2019).

Andrea Wright is a Senior Lecturer at Edge Hill University, UK. Fantasy/fairy tales, New Zealand cinema, and television costume drama are central to her research interests. She has published scholarship on production design, landscape, gender representation, and national identity.

Index

Abraham, Nicolas 127
Abrams, Lynn, "Ideals of womanhood in Victorian Britain" 183 n.47
abstract music 63–4
acousmatic sound 42, 49, 131, 154
acoustics 28, 49, 92, 96, 98, 106, 112, 118, 124, 131, 160, 177, 190
   acoustic ecology 105, 123, 130
   acoustic ectoplasm 28–9, 31, 33, 36, 37 n.10
   acoustic instruments 50–1, 60, 154–6
   electroacoustic music 93–4
   myth of acousticity 153, 160
Adorno, Theodor 46, 166
aesthetics 4, 10–11, 13, 18, 21, 27–8, 32, 42–3, 89–90, 108, 155, 169, 174, 176, 180 n.10
alien/alienation 4, 10, 59, 114, 151–2, 155–6, 159–61, 165, 173, 175, 186–7, 189–90
ambient sound 12–13, 16, 18, 30–1, 33, 112, 116, 118
ambiguity/ambiguous 3–5, 27–9, 32–3, 41, 43–4, 49, 52, 61–4, 68, 86, 91, 101, 137, 144, 160, 165–6, 168, 178, 180 n.6, 189, 191
analogue instrumentation 153–4
analogue technology 4, 6, 151, 153
Anderson, Benedict, imagined community 175
anempathetic sound 15
Annihilation (Alex Garland) 6, 151–2
   alien/alienation 151–3, 155–6, 159–61
   "All along the watchtower" 154–6
   Anya Thorensen (fictional character) 158, 161
   bear (Homerton) scene 162 n.3
   Cass Sheppard (fictional character) 158
   "Cells Divide" soundtrack 156–7
   Dan (fictional character) 157–8
   Dr. Ventress (fictional character) 158–9
   folk/blues mythos 156, 161

"Helplessly Hoping" song 154, 157–9
Josie Radek (fictional character) 158
Kane (fictional character) 156–8, 160, 163 n.18
Lena (fictional character) 156–61
Lomax (fictional character) 161
music and analogue instrumentation 152–4
self-destruction 151–3, 156–8
shimmer 151, 156, 158–60
Anthropocene 108–9, 116
Antipodean 186, 195. *See also* Australia; New Zealand
Antonioni, Michelangelo
   Blow-Up (*see Blow-Up* (Michelangelo Antonioni))
   Houston on 11
anxiety(ies) 1, 6, 41, 43, 46–7, 53, 67, 165, 174, 186, 194
Arctic Noir. *See also The North Water; The Terror: Season 1*
   sonic extinction 109–18
   spectrality of 106–9, 116
   wind (windscape) sounds 106, 109, 112–13, 115–18
Arrowsmith, William 11
Artaud, Antonin 145
artifacts 2, 4, 73, 75–6, 79–82, 84, 194
audiovisual culture 1–3, 5, 10, 34, 36, 41–4, 46–7, 52, 90–1, 97, 123, 136–8, 157, 166–71
auditory nostalgia 129
aural/aurality 2, 27, 36, 109, 117–18, 129, 185, 190, 193–5
Auric, Georges 59
Australia 6–7, 165–6, 174. *See also Picnic at Hanging Rock* (Peter Weir)
   Aboriginal people 179
   Australia Knowledge Base 180 n.9
   Gothic film genre (Australian Gothic Cinema) 168–71, 179
   New Wave cinema movement 165, 176–8

authentic/authenticity 32, 49, 51, 63, 105, 115, 153, 155, 160, 175–6
authorial control 9, 20–1
autonomy 9, 20, 27, 44, 52
avantfolk 155, 162 n.13
avant-garde 52, 67

The Babadook (Jennifer Kent) 140–2
    Amelia (fictional character) 140, 147–8 n.20
    Babadook (fictional character) 140, 147 n.20
    Samuel (fictional character) 140, 147–8 n.20
Bachelard, Gaston 4–5, 31, 35
    and Maddin 26–7, 32, 36, 37 nn.3–4
    The Poetics of Space 26–7, 32, 36, 37 n.3, 39 n.25
    reverberation 27, 29, 31
    topophilia 32
Bakhtin, Mikhail, literary chronotope 4, 10
Balázs, Béla 54 n.10
Banks, Joe, on EVP recordings 62–3
Barrow, Geoff 153
BBC 5, 57–8, 67, 73, 82, 105
BBC Radiophonic Workshop 5, 57, 67, 69 n.4, 71 n.25
    Cubitt on 68
    and electronic music 58–9
Beethoven, Ludwig van
    "Emperor" concerto 167
    "Moonlight Sonata" 92
    Piano Concerto No.5 166–7, 169
Belasco 60, 67
Bell, Claudia 185
Blood on Satan's Claw (Piers Haggard) 152
Blow-Up (Michelangelo Antonioni) 4, 20–2
    Bill (fictional character) 18
    honour for 9
    individuals worked on film sound 22 n.4
    Jane (fictional character) 15–16, 21
    shot in Maryon Park (sounds of leaves blowing in trees) 9–20
    Thomas (fictional character) 4, 9–16, 18–22

Bordwell, David, *Making Meaning: Inference and Rhetoric in the Interpretation of Cinema* 43
Born, Georgia 17
Bowers, Katherine 106–7, 116–17. *See also* polar Gothic/Gothicism
    on "Rime of the Ancient Mariner" (Coleridge) 109
Boym, Svetlana, on nostalgia 125, 128
Briscoe, Desmond 57, 60, 69 n.4
    Children of the Damned 58
    The Haunting 58–9
    The Ipcress File 58
    The Man Who Fell to Earth 58
British historical dramas 123. *See also Sunset Song*; *Wuthering Heights*
    mythological landscapes and trauma 124–6
    Romantic 125, 131
    temporality and sonic haunting 130–1
    trace and hauntology 127–30
British television 5, 59
Brooker, Peter, on Gothic genre 169, 181 n.17
Brunette, Peter 12
Bubandt, Nils, *Arts of Living on a Damaged Planet: Ghosts and Monsters of the Anthropocene* 116
Buhler, Jim 91

Campion, Jane 194
Carroll, Beth 28–9
Catania, Saviour 180 n.6
Ceuterick, Maud 188
Chion, Michel 11–12, 95–6, 169
    anempathetic sound 15
    reduced listening 11
    rendered sound 95
chronotopes 4, 10
cinematic soundscapes 105
cinematography 93, 95, 99
    and soundscape 99–100
classical music 7, 141
    in *Picnic at Hanging Rock* (Peter Weir) 166–71, 176, 178–9
climate catastrophe 106, 108, 111, 116
Coates, Paul, German Expressionist Cinema 191

cognitive dissonance 154–5
Coleridge, Samuel Taylor, "Rime of the Ancient Mariner" 109
collective memories 3, 29, 123–5. *See also* memory(ies)
colonial/colonialism 6–7, 107, 109, 111, 116–18, 165–6, 170, 174, 178–9, 187
colonialization 170–1
colonization 53, 106, 165, 167, 179, 193
concrete sounds (diegetic sounds) 97
Configurationism. *See Gestaltist* approach
Connell, John, music tourism 175
Conradh na Gaeilge (the Gaelic League) 124
Conrich, Ian 185–6, 194
constructed sounds 10–11, 105
conventional music 27, 49, 53, 64, 69 n.1, 90, 96, 166–71, 179
Cubitt, Kirsten, on Radiophonic Workshop 68
cultural geography 3, 123
Curtis-Bramwell, Roy 69 n.4

Davison, Annette 96
dead sound 48, 57, 62, 105
death 6, 26, 41, 46, 106, 108, 110–11, 116, 137–46
Debord, Guy 4
Delaware. *See* EMS Synthi synthesizer
Deleuze, Gilles 161
  refrain 131
demonic force 135, 140
Derbyshire, Delia 57–8, 69 n.2, 69 n.4, 70 n.5
Derrida, Jacques 2, 4, 17, 161
  on cinema 128
  Echographies of Television 129
  on hauntology 3–4, 152
  specter 18, 20, 129–30
  Specters of Marx 43, 127–8
  spectral visuality 129
Devs (Alex Garland) 153
dichotomy 7, 91–2, 154, 160, 177, 180 n.10
diegetic music 5, 16, 18, 28–9, 43, 49, 65, 89–92, 94–7, 103 n.22, 138, 140–1, 143, 154, 168, 173, 189. *See also* non-diegetic music; Van Sant, Gus, *Death Trilogy*

Digital Audio Workstations (DAWs) 153–4
digital technology 6, 30, 44, 47
disc 71 n.18
dissonant sounds 60, 65, 67, 154
distant sounds 48, 142–3
Distant Voices, Still Lives (Terence Davies) 126
Django Unchained (Quentin Tarantino) 111–12
Dolgopolov, Greg 191–2
domestic objects 28, 30
domestic space 7, 27, 31, 185–6, 190–1, 195
Donnelly, K. J. 71 n.24, 92, 135, 140–1, 154
  on integrated soundtracks 65
  The Spectre of Sound: Film and Television Music 44, 135
doppelgängers (doubles) 6, 152, 157, 160, 162 n.17
Doyle, Christopher 99
Doyle, Peter, on echo and reverb 173–4
Dreams within a Dream (Peter Weir) 180 n.6
drones 31–2, 41, 49–51, 60–1, 64, 98–9, 143, 147 n.20
Duncan, James, on landscape 1–2, 47
Duncan, Nancy 1
Duncan, Paul 21
DVD format 38 n.14, 147 n.14
  DVD Extras 146 n.7

Echographies of Television (Jacques Derrida) 129
ecology 2, 16, 107–9, 112, 118
  acoustic 105, 123, 130
  ecological catastrophe 106, 108
ectoplasm. *See* acoustic ectoplasm
Eine Kleine Nachtmusik: Romance (Mozart) 167
Eisenstein, Sergei
  nonindifferent nature 45–6
  talking films 45, 54 n.9
Eisler, Hanns 46, 166
electro-acoustic music 90, 93–4, 130
electronic music 57, 60, 64, 68, 112
  ambient 118
  BBC Radiophonic Workshop and 58–9
  techno ambient 112

electronic scores 57–9, 65, 69 n.1, 70 n.7, 84
electronic sounds 5, 46, 49, 59, 64, 67–8, 70 n.7
  experimental 64, 68
  in *The Legend of Hell House* (see *The Legend of Hell House* (John Hough))
  in *The Stone Tape* (see *The Stone Tape* (Peter Sasdy))
electronic tonalities 59
Electronic Voice Phenomena (EVP) 57, 61–2, 70 n.18
  Banks on EVP recordings 62–3
Electrophon Studios 57–8, 69 n.2
Elephant (Gus Van Sant) 89–94, 97, 101
  Alex (fictional character) 90, 93, 98–100
  Eric (fictional character) 100
  John (fictional character) 97
  Macy (fictional character) 93
  Nathan (fictional character) 92
emergent perception of sound 16–17
emotions 3, 26–7, 29, 32, 45, 49, 74, 80, 82, 95–7, 99, 103 n.22, 138, 142, 174–5
  emotional authenticity 175
  emotional landscape shots 45–6
  emotional representation 45
  emotional sensitivity 68
EMS Synthi synthesizer 58, 67, 69 n.4
EMS VCS 3 synthesizer 58, 69 n.4
enigma 165–6, 169, 171, 173–4, 176, 178–9
environmentally induced distress 108
environmental sounds 11, 17, 105, 190
Eraserhead (David Lynch) 27
Erebus & Terror (David Bickley and Tom Green) 112
European 7, 47, 123, 165, 171, 174, 179, 186
Ex Machina (Alex Garland) 153
exotic/exoticism 7, 55 n.21, 59, 166–79, 180 n.10
extinct sounds 105–6, 112
extra-diegetic music 16, 18, 141, 148 n.27. *See also* intra-diegetic music
extraneous sound 50, 62
extra-sensory perception (ESP) 62, 68

fabricated landscape 2, 4, 124
Fahey, John 154–5
  "Dance of Death" 155
The Fall of the House of Usher (Edgar Allen Poe) 190
fantastical gap (diegetic and non-diegetic) 5, 91
felicitous space 32
feminine 176, 185, 189
feminism in cinema 176–8
fetishism 175, 180 n. 10
A Field in England (Ben Wheatley) 152
film(s). *See specific films*
filmind 29, 135–7, 140–2, 144–5, 148 n.27
film noir 25. *See also Keyhole* (Guy Maddin)
film sound 10, 13, 22 n.4, 27, 96
Fisher, Mark 43, 82, 128–9
  on hauntology 17–18, 108
  on weird 154, 160
Fjellström, Marcus 112, 114
flexidisc 71 n.18
Flûte de Pan (pan flute/syrinx) 173
folk horror films 4, 47, 151–2. *See also Annihilation* (Alex Garland); horror films; *The VVitch* (Robert Eggers)
folk music 6, 77, 136, 154–6, 159, 175
The Forbidden Planet 59
The Forbidden Room (Guy Maddin) 25
Foreman, Iain 21
Frampton, Daniel
  filmind (see filmind)
  Filmosophy 38 n.13
fluid filmind 29
Franklin expedition 107, 110–11
Freud, Sigmund 191
friction ideophone 51

Gan, Elaine, *Arts of Living on a Damaged Planet: Ghosts and Monsters of the Anthropocene* 116
Garner, Alan
  "Inner Time" 74, 83
  The Owl Service (see *The Owl Service* (Alan Garner))
  Red Shift (See *Red Shift* (Alan Garner))
Garner, Tom 16
Gassmann, Remi 70 n.7

gender 68, 186, 191–2, 195
geography 46, 123
   cultural 3, 123
   geographic locales 123, 131
   psychogeography 3–4, 6, 28, 132, 159
   sonic 6, 106
geophones 2
Gestaltist approach 167, 174, 176, 179
Ghost Dance (Ken McMullen) 128
ghosts/ghost stories 1, 3, 5, 16, 25–32, 43–6, 49, 57, 59, 60, 62–6, 68, 73–4, 86, 105, 108, 116, 124, 127–9, 154, 160. *See also Keyhole;* specter/spectrality
   Ghostly Effect/reverse Ghostly Effect 46
   Lethbridge on 62
   of trauma 135
Gibson, Chris, music tourism 175
Gibson, Ross 168, 170
Gibson, Suzie 180 n.6
Goodman, Steve 67
Gordon, Assheton 10
Gothic architecture 71 n.27
Gothic/Gothic horror 6, 65–7, 168–70, 179, 181 n.17, 187, 190, 195
   van Elferen on sounds of 60
Grace, Sherrill 107
Graham, Davy 154
grief 6, 135, 137, 140, 142–3, 145, 148 n.31, 158
Griffith, D. W. 9
Grimshaw, Mark 16
Guattari, Felix, refrain 131
Guilbault, Jocelyne, on world music 175
Gunning, Tom 46
Gurdebeke, John 29–30, 32
   "Glorious Cut 2" 32
   "Glorious Cut #7" 33, 38 n.21

Happiness (Adam P. Wadenius) 192
haptic soundtrack 37 n.11, 136
harmony 28, 34, 36, 51–2, 142, 144, 160
Harper, Don 58
harp music 77–80
haunted houses 1, 28–9, 31, 58, 60, 66, 194
The Haunting (Desmond Briscoe) 58–9

hauntings 2–4, 17–18, 53, 67–9, 80, 105–9, 128, 136, 151, 154, 194
   of Arctic Noir 109–18
hauntology 43–4, 123, 130
   Derrida on 3–4, 152
   Fisher on 17–18, 108
   trace and 127–30
Hecker, Tim 112–13
hi-fi soundscape 16
high-fidelity recordings 63. *See also* low-fidelity recordings
historical dramas. *See* British historical dramas
historical soundscape 2, 82, 105–6, 112, 115, 118
Hitchcock, Alfred, *The Birds* 70 n.7
Hodgson, Brian 57–8, 69 n.2, 69 n.4, 71 n.30
homonym ontology 127
horror films 28, 46, 49, 52–3, 57, 59–61, 67–9, 109–10, 140, 145, 147 n.20, 162 n.3, 168–71, 188, 193. *See also* folk horror films
Houston, Penelope 11
Hugh, Soo 115
humanity 6, 151, 153, 155, 192
hurdy-gurdy (bowed string instrument) 50
Hutchings, Peter 46
hydrophones 2, 115
hyperreal sounds 93, 124, 131

imagined soundtracks 175–6
imperialism 107, 111, 116, 174, 178
incidental music 43, 49, 167
industrialization 111, 118
The Innocents (Daphne Oram) 59
intra-diegetic music 102 n.15, 140–1, 148 n.27. *See also* extra-diegetic music

Jansch, Bert 154
jazz music 18, 20, 55 n.21
Jones, Glynis 57, 61

Kassabian, Anahid 176
   Hearing Film 90–1
Kavka, Misha 193–4

Keyhole (Guy Maddin) 5, 25–8
  Calypso (fictional character) 26, 35
  Denny (fictional character) 25, 30–2
  Hyacinth (fictional character) 25–6, 30–3
  Lota (fictional character) 31
  Manners (fictional character) 25, 30–3, 35–6
  review of 37 n.3
  soundtrack of 27–36, 38 nn.15–16
  Ulysses Pick (fictional character) 25–6, 28, 30–3
keynote sounds 131, 133 n.31
Kibby, Marjorie D., "Great Australian Silence" 171–2
Kingsland, Paddy, "Scene and Heard" 69 n.1
Kirmayer, Laurence 125, 132
Kiwi Gothic 185–6
Kneale, Nigel 70 n.15
Kohn, Eduardo 16
Korven, Mark 44, 49–51, 53, 55 n.21
KPM library 58
Kracauer, Siegfried 9, 20
Krause, Bernie, totem soundtracks 131
Kurzel, Jed 135, 146 n.7, 147 n.16, 149 n.34. *See also Macbeth* (Justin Kurzel)

Lady Macbeth of the Mtsensk District novel (Nikolai Leskov) 123
landscape(s) 1–7, 11, 41, 105–6, 123, 146 n.8, 170, 185
  Arctic (uncanny) 106–18
  British rural landscape 6, 123
  disturbing landscape in *Picnic at Hanging Rock* 172
  Duncan (James) on landscape 1–2, 47
  fabricated 2, 4, 124, 131
  haunted/hauntological 17, 46–8, 80, 86, 124, 127, 129, 131–2, 155, 159–60, 166, 187
  mythological 124–7, 130
  narrative 125
  romanticized 185
  and sound 73–5 (*see also The Owl Service* (Alan Garner); *Red Shift* (Alan Garner))
  stratified 5, 75, 77, 82, 84, 86

  symbolic 123–4
  of trauma 136–40
La Salle Terre novel (Honoré de Balzac) 123
Last Days (Gus Van Sant) 89
  Blake (fictional character) 90, 102 n.4
The Last Wave (Peter Weir) 165
Law of Closure 167, 179
Law of *Prägnanz* 167, 176
Leeder, Murray 29, 61
The Legend of Hell House (John Hough) 5, 57, 68–9, 69 n.1
  ambiguity, sound, and projection 61–4
  Dr. Barrett (fictional character) 65, 67
  electronic sounds in 59–61
  influence of spiritualism on 61
  scientific machine 65, 67, 71 n.26
  sound effects and music 65–7
Le Grand Blond avec une Chaussure Noire (Gheorghe Zamfir) 182 n.29
Leotta, Alfio 186, 189
Lethbridge, T. C. 57, 61, 68, 70 n.15
  on ghosts 62
Lewis, John 69 n.2
Lippit, Akira Mizuta 128–9
literary chronotope 4, 10
Lively, Penelope 73, 81
living space 26. *See also Keyhole* (Guy Maddin)
London Contemporary Orchestra 146 n.10
loop-based motifs 61. *See also* motifs
Loscil, "Goat Mountain" 112
low-fidelity recordings 62–4. *See also* high-fidelity recordings
low-frequency sounds 60, 114, 118
Lucier, Alvin, "I Am Sitting in a Room" 63
Lux Aeterna (György Ligeti) 52
Lynch, David 103 n.22
  Eraserhead 27

Macbeth (Justin Kurzel) 6, 135, 141, 148 nn.31–2
  Banquo (fictional character) 141, 143
  DVD release of 147 n.14
  filmind 140–2
  King Duncan (fictional character) 137–40, 142–3

Lady Macbeth (fictional character) 137–8, 143–5
landscape of trauma 136–40, 142
Macbeth (fictional character) 137–46
Macduff (fictional character) 143–4
musicians from London Contemporary Orchestra 146–7 n.10
post-traumatic stress disorder (PTSD) 137, 139
self-destruction 135, 145
slow-motion shots and muted diegetic sound 137–9, 141
supernatural force 136–7, 139, 146
trajectory of death 145–6
witches 136–41, 144–5
Mackenzie, John 82
mainstream music 169, 174–5, 178. See also non-mainstream music
Māori 187. See also Pakeha
"The Mark," Moderat 160
masculine/masculinity 52, 186, 189, 192–3
materiality 6, 46–9, 63, 135–6
materialization 28, 30, 33
McCorristine, Shane, *The Spectral Arctic: A History of Dreams and Ghosts in Polar Exploration* 107, 110–11, 116–17
Meinig, Donald W. 123
melancholic/melancholy 109, 113, 136, 173
melody/melodic 27, 29–31, 33, 50–1, 140, 142, 144
memory(ies) 1–3, 6, 25, 28–9, 31–2, 43–4, 74, 117, 123, 125, 130–1, 138–9, 145, 151–2, 158, 160, 174, 194. See also collective memories
Men (Alex Garland) 152–3
meta-diegetic space 141–3
meta-human 10, 13, 20–1
micropolyphony 52
microtonal music 32, 136, 139, 140, 142, 154
Midsommar (Bobby Krlic) 55 n.22
Milner, Johnny, imagined soundtracks 175–6
Minimoog synthesizer 69 n.4
Molloy (Samuel Beckett) 159

monster movie. See *The Babadook* (Jennifer Kent)
Moog synthesizer 69 n.2
Morgan, Frances 69 n.4, 71 n.30
Morton, Timothy 108
motifs 60–1, 77–8, 81, 113–14, 144–5, 147 n.20
moving images 3–5, 9, 46
Mulvey, Laura, male gaze 176
Murch, Walter 101
music tourism 175
musique concrète 5, 58, 66, 89–101, 155. See also Van Sant, Gus, *Death Trilogy*
mystery/mysterious 12–14, 16, 19–20, 65, 68, 107–9, 130, 136, 138, 153, 185–8, 195. See also *Picnic at Hanging Rock* (Peter Weir)
mythology 48, 94, 123, 156, 173, 185, 193
mythological landscapes 124–7, 130

Nancy, Jean Luc, "Uncanny Landscape" 107
narrative space 92, 137–8, 141, 148 n.27
Narváez, Peter, myth of acousticity 153, 160
nature 2, 7, 10, 21, 45, 107, 124, 168, 170–1, 185–7, 189–91, 193–5
 natural environment 1, 13, 123, 127, 166, 185, 190
 natural sounds 10, 20, 131
Nature's Way (Jane Shearer) short film 7, 185, 187–92, 194–5
 Yvonne (fictional character) 190, 192–3
neo-colonialism 175
neologism 38 n.13
New Wave cinema movement 165, 176–8
The New World (Terence Malick) 105–6
New Zealand 6–7, 185–6
 Aotearoa New Zealand 187
 New Zealand Film Commission's Short Film Fund 186
 Pakeha (white New Zealanders of European decent) 186
 settlement/settler anxieties 185–8, 193–4
Niebur, Louis 58, 61, 67, 69 n.4

non-diegetic music 5, 44–5, 48–9, 65, 89–92, 94, 135, 140, 154, 156, 188, 190–1, 194. *See also* diegetic music; Van Sant, Gus, *Death Trilogy*
non-electronic sound 60, 67
nonhuman sound 16–17, 20
nonindifferent nature 45–6
nonlinear music 60
non-mainstream music 174, 178. *See also* mainstream music
Normandeu, Robert, *La Chambre Blanche* 95
The North Water novel (Ian McGuire) 110–11, 117
The North Water television series 6, 106, 109–10, 112–13
   Brownlee (fictional character) 114, 117
   CGI 111–12
   Henry Drax (fictional character) 110–11, 113–14, 117
   "Homo Homini Lupus" (episode 3) 113–14
   Joseph Hannah (fictional character) 110, 114
   Otto (fictional character) 113
   Patrick Sumner (fictional character) 110–11, 113–14
   "We Men Are Wretched Things" (episode 2) 113
   whale-like elements 112–14, 116, 118
nostalgia 1–2, 105, 116, 124–5, 128–9, 131, 152
   auditory 129
   Boym on 125, 128
nyckelharpa (keyed violin) 50–1

occult 4–5, 25, 31, 35, 57, 61, 65, 174
off-screen sound 79–81, 85–6, 100
Olenick, Mike 38 n.18
orchestral music 50, 59
organic sound 48–9, 51
The Owl Service (Alan Garner) 5, 73–81
   Alison (fictional character) 75–6, 79–81
   Bertram (fictional character) 76, 80–1
   fiming location 76
   ghosts in 77, 81

   Gwyn (fictional character) 75–6, 79–81
   harp music 77–80
   Huw (fictional character) 76
   The Mabinogion stories 75–6, 80–1
   mythic time 75–9
   Nancy (fictional character) 76
   Roger (fictional character) 75–7, 79–80
   scratching sound 73, 77–8, 80, 86
   "Ton Alarch" song 77

Pakeha (white New Zealanders of European decent) 186–7
panpipe music 7
   in *Picnic at Hanging Rock* (Peter Weir) 166, 171–9
Paranoid Park (Gus Van Sant) 5, 89–101
   adjacent bubbles 91
   Alex (fictional character) 90, 93–8, 100–1
   Detective Liu (fictional character) 97, 100
   diegetic/non-diegetic sounds 94–7
   electro-acoustic music in 90, 93–4
   Jared (fictional character) 96–7
   Scratch (fictional character) 100
   "Song One" (Ethan Rose) 93–6
   "Song Three" 100
   Tommy (fictional character) 93
   Walk through Resonant Landscape No.2 (Frances White) 93–5, 97–8
paranormal 5, 46, 53, 57, 61, 65
parapsychology 28
Paravisini-Gebert, Lizabeth 187
patriarchal society 176–7, 191, 193
Payne, Robert, *Songs of the Humpback Whale* 109
Penda's Fen (Alan Clarke) 82
period dramas/films 6, 105, 112, 125, 128
phantoms 2, 107, 127–9
Phase IV (Saul Bass) 58
phonotope 4, 10–11, 20, 22 n.10
Picnic at Hanging Rock (Peter Weir) 6, 165–6
   classical music in 166–71, 176, 178–9
   disturbing landscape in 172
   Doina Lui Petru Unc (n.d.) 173
   Doina: Sus Pe Culmea Dealului (n.d.) 173

feminism (Other/Otherness) 176–8
girl appears and disappears in cave 169–70
New Wave cinema movement 165, 176–8
panpipe music in 166, 171–9
restriction, symbolism of 178
Sara's corpse in the garden 171
soundtrack ranking 180 n.10
Pierce, Peter, on lost child in Australian cinema 165
Pieterson, Daniel, spectral resonance 63
pipe organ musical instrument 60
Play for Today series (BBC) 73, 82
Plummer, Peter 76
Polar Gates (Ugasanie and Dronny Darko) 112
polar Gothic/Gothicism 6, 106–7, 109, 112–13, 116–18
Ponti, Carlo 9
Possum (Brad McGann) short film 7, 185, 187–9, 191–5
　animality 188–9, 191–2
　Kid (fictional character) 188–9, 192–4
　Little Man (fictional character) 188–9, 193
　Missy (fictional character) 191–2, 194
　porous environment 188–9
pseudo-authenticity 175–6
pseudonyms 58, 70 n.5
psychogeography 3–4, 6, 28, 159

Quatermass and the Pit (BBC) 152

Ratner, Megan, "Paranoid Park: The Home Front" 99–100
Raudive, Konsantin, *Unhörbares Wird Hörbar* (The Inaudible Becomes Audible) 62
recorded sound 11, 13, 21, 50, 57–8, 63, 74, 115, 173
recording process 6, 44, 46, 51, 62–3, 71 n.18, 78, 106, 115, 118, 130
recreating sounds 6, 105, 115
Red Shift (Alan Garner) 5, 73, 75, 81–6
　editing and sound 81, 83–5
　Jan (fictional character) 83–5
　Macey (fictional character) 82–6

Madge (fictional character) 84
Thomas (fictional character) 82–5
Tom (fictional character) 82–5
refrain 131
Requiem (György Ligeti) 52
re-recording process 63–4
The Revenant (Alejandro González Iñárritu) 111
reverberation 26–7, 29, 31, 63, 78, 113, 173–4, 189–90, 195
Reyes, Xavier Aldana 168
Reynolds, Simon 43, 152
rhythm 14–15, 27, 30, 33, 51, 58, 137, 144, 189
Romantics 125, 145, 152
Royle, Nicolas 187, 191
Rudkin, David, *Penda's Fen* 82
Ruffles, Tom, *Ghost Images: Cinema of the Afterlife* 46
Ryan, Phil 84

Said, Edward 124–5
　"Invention, Memory, and Place" 123
Sala, Oskar 70 n.7
Salisbury, Ben 153, 155
Save Our Sounds, BBC's 105
Savides, Harris 89
Schafer, R. Murray 2, 112, 130
　hi-fi soundscape 16
　keynote sounds 131, 133 n.31
　The Soundscape: Our Sonic Environment and the Tuning of the World 105, 133 n.31
Schafer, William J. 187
Schell, Jonathan, on temporality of postwar era 128
science 5, 26, 57, 61, 65
science fiction 6, 57, 59–60, 68, 151
scientific horror 68
Sconce, Jeffrey, *Haunted Media: Electronic Presence from Telegraphy to Television* 46
screen media 5, 36, 59
Séances project (Guy Maddin) 25
Searle, Humphrey 59
sexuality 53, 192
Shakespeare, William 135, 137, 141, 148 n.32

Shatz, Leslie 89
The Shining (Stanley Kubrick) 55 n.27
short films 7. *See also Nature's Way;
    Possum*
Silence (Pat Collins) 125
Sinclair, Iain, on Maryon Park 9
SleepResearch_Facility, *Deep Frieze* 112
slow-motion shots 94, 137–9, 141, 145, 165
Snowtown (Justin Kurzel) 148 n.31
solastagia 107–9, 118
sonic aggregate 16
sonic assaults 64, 99
sonic chiaroscuro 190
sonic conventions 59, 109
sonic extinction 105–6, 109
    hauntings of Arctic noir 109–18
sonic haunting 1, 5–6, 73, 90, 112, 114,
        123, 154
    in British historical dramas 123
    temporality and 130–1
sonic spaces 4–5, 7, 13–14, 89–90, 94
sonic trope 60, 115
sonic turn 3
Sonnenschein, David 97
sound continuum 4, 27, 52–3
sound design 5, 11, 27, 76, 85, 94, 142,
        160, 195
soundscape 2–3, 7, 21, 32, 52, 89–90,
        93–5, 99–102, 105, 118, 130,
        174, 185, 195
    cinematography and 99–100
    hi-fi 16
    historical 2, 82, 105–6, 112, 115, 118
sound space 2, 4
    of *Blow-Up* film 9–20
soundtracks 1, 3–5, 11, 20, 30, 65–7, 77,
        81, 85–6, 91–3, 95–6, 106, 109,
        112, 151, 154–6, 166, 173, 178,
        180 n.9, 188–90
    cosmic-folk 151
    electronic 5, 59–60
    imagined soundtracks 175–6
    of *Keyhole* film 27–36, 38 nn.15–16
    totem soundtracks 131
    of *The VVitch* film 41–53
Spadoni, Robert, *Uncanny Bodies: The
    Coming of Sound Film and the
    Origins of the Horror Genre* 46

spatial listening 11
Specters of Marx (Jacques Derrida) 43,
    127–8
specter/spectrality 2, 18, 20, 46, 69, 91,
    101, 106, 127–30, 151, 154–5,
    159, 191. *See also* ghosts/ghost
    stories
    of Arctic soundscapes 106–9, 116
    spectral presence 1–3, 6, 17, 28–9,
        35–6, 127, 129–30, 132, 135
    Toop on spectral nature of sound
        17–18
spiritual/spiritualism 28, 61, 107, 174, 177
    on *The Legend of Hell House*, influence
        of 61
Staczek, Jason 27, 38 nn.14–16
"Shaft" 29–32, 38 n.15
Standard Music Library 58
Stilwell, Robynn 91
The Stone Tape (Peter Sasdy) 5, 57–8,
    68–9, 69 n.1
    ambiguity, sound, and projection 61–4
    electronic sounds in 59–61
    Jill (fictional character) 63, 65–6, 68,
        72 n.33
    Peter (fictional character) 68, 72 n.33
    sound effects and music 65–7
Sunset Song film (Terences Davies) 6, 124,
    126–9, 132
    Chris Guthrie (fictional character) 126,
        129, 131
    Ewan (fictional character) 126, 129
Sunset Song novel (Lewis Grassic Gibbon)
    123, 126
supernatural/supernaturalism 4, 6, 10, 16,
    44, 46–7, 49, 52–3, 54 n.5, 61,
    68, 106–7, 109–11, 115, 118,
    123–4, 136–7, 139, 146, 186, 192
Swanson, Heather, *Arts of Living on a
    Damaged Planet: Ghosts and
    Monsters of the Anthropocene* 116
synchronicity 83, 92, 98, 154, 157
synthesizers 58–9, 66–7, 69 n.4
synthetic sounds 61, 166

tape editing 58
Tarr, Béla 89
Tattersall, Jane 113

Tayfora, Scout 100
Taylor, Timothy D. 174-5
techno ambient electronic music 112
technostalgia 151, 153
television 1, 3, 58, 73-4, 86, 105, 112, 186, 190
temp music 32, 38 n.15
temporal/temporality 2, 4-5, 32, 58, 74, 76, 85-6, 128-9
  Schell on temporality of postwar era 128
  and sonic haunting 130-1
The Terror novel (Dan Simmons) 110-11
The Terror: Season 1 television series 6, 106, 109-12, 114-16
  CGI 111-12, 115
  Crozier (fictional character) 110
  live recordings in snow 115
  "Terror Camp Clear" (episode 15) 117
  Tuunbaq 110, 115-16
Thacker, Eugene 69
theremin (electronic instrument) 59
timbres 41, 57, 60, 64, 113, 115
time-space 10, 17-18, 21, 46, 63, 74, 78, 80-1, 86, 91, 97
To Kill a Mockingbird (Robert Mulligan) 34-6, 39 n.23
Toles, George, "Being Well-Lost in Film" 34-5
The Tomorrow People series 59
Toop, David, on spectral nature of sound 17-18
Torgue, Henry 64
Torok, Maria 127
totem soundtracks 131
transgenerational communication 127
trauma/traumatic events 1-3, 5-6, 31-2, 63, 99, 114, 123, 127-9, 131, 135, 152, 159
  in British historical dramas 124-6, 128
  Kirmayer on 132
  landscapes of (*Macbeth*) 136-40, 142, 144-5
Tsing, Anna, *Arts of Living on a Damaged Planet: Ghosts and Monsters of the Anthropocene* 116
Twilight Zone (Bernard Herrmann) 32
2001: A Space Odyssey (Stanley Kubrick) 52, 155

van Elferen, Isabella 60
  "Dream Timbre: Notes on Lynchian Sound Design" 103 n.22
  on film music 169
Van Sant, Gus, *Death Trilogy*. See also *musique concrète*
  Death Quartet 5, 89-92, 96, 101
  diegetic and non-diegetic music 89-92, 94
  Elephant (see *Elephant* (Gus Van Sant))
  Gerry 89
  Last Days (see *Last Days* (Gus Van Sant))
  Paranoid Park (see *Paranoid Park* (Gus Van Sant))
vinyl 18, 62, 71 n.18, 153
violence 6, 84, 109, 114, 116, 125-6, 131, 139, 142, 145, 146 n.8, 192
visibility/invisibility of image 2, 20, 48, 129
vocal sounds 52, 66
Volk identity 124
Vorhaus, David 58
The VVitch (Robert Eggers) 4, 41
  Black Philip (fictional character) 42, 52
  Caleb (fictional character) 41-2, 52
  emotional landscape shots 45-6
  ghosts and nonindifferent nature 45-6
  haunted landscape in 46-8
  Jonas (fictional character) 41-2
  Mercy (fictional character) 41-2
  Samuel (fictional character) 47-8
  sound and image 42-4
  Thomasin (fictional character) 41-2, 48, 53
  Walton on 54 n.5
  William (fictional character) 41-2, 47, 51
  wooden sounds 48-52

Wake in Fright (Ted Kotcheff) 165
Walpole, Lee 115-16
Walton, Saige 37 n.11, 54 n.5, 54 n.10
Ward, Vincent 194
waterphone 51, 153-5

weird/weirdness 6, 151–2, 154–5, 159–60, 175
The Well-Tempered Clavier (Bach's Prelude No.1 in C major) 166
Wertheimer, Max, on *Gestalt* theory 167
Westerkamp, Hildegard, *Doors of Perception* 92
Western Soundscape Archive 105
White, Frances, *Walk through Resonant Landscape No. 2* 93–5, 97–8
Whitehead, Alfred 17
Winters, Ben 92, 140–1, 148 n.27
    intra-diegetic music 102 n.15, 140–1
    "The Non-Diegetic Fallacy" 102 n.15
world music 174–6
    Guilbault on 175
The World Soundscape Project 105

Wuthering Heights film (Andrea Arnold) 6, 124–5, 127–30, 132
    Cathy (fictional character) 125–6, 129–31
    Edgar Lynton (fictional character) 126
    Hareton (fictional character) 125
    Heathcliff (fictional character) 125–6, 130
    Isabella (fictional character) 125–6
    keynote sounds in 131
    violence 125–6
Wuthering Heights novel (Emily Brontë) 123

Zamfir, Gheorghe 173
    Le Grand Blond avec une Chaussure Noire 182 n.29

www.ingramcontent.com/pod-product-compliance
Lightning Source LLC
Chambersburg PA
CBHW052041300426
44117CB00012B/1920